EMPOWERING LASER TECHNOLOGY WITH MACHINE LEARNING

Tayyab Imran

Machine Learning Insight Into The Future of Laser Technology

Empowering Laser Technology with Machine Learning

Machine Learning Insight Into The Future of Laser Technology

Tayyab Imran, M.Phil., Ph.D.
Extreme Light Infrastructure-Nuclear Physics (ELI-NP),
'Horia Hulubei' National R&D Institute for Physics and
Nuclear Engineering (IFIN-HH), Magurele, Romania

Copyright © 2024 Tayyab Imran

All rights reserved

No part of this book may be reproduced, or stored in a retrieval system, or transmitted in any form or by any means, electronic, mechanical, photocopying, recording, or otherwise, without express written permission of the publisher.

ISBN-13: 9798346955481

Cover design by: Tayyab Imran
Library of Congress Control Number: 2018675309
Printed in the United States of America

DEDICATION

To Allah, for His endless guidance and blessings.
To my wife, for her love and support, and to my son, for bringing joy
and purpose to my life.

CONTENTS

	Preface	1
	Introduction	4
1	Overview of Laser Technology	7
2	The Journey of Machine Learning	16
3	Artificial Intelligence and Machine Learning Hand In Hand	20
4	Basics of Machine Learning	25
5	Intersection of Laser Technology and Machine Learning	37
6	Machine Learning Techniques in Laser Applications	43
7	Advanced Laser Technologies	54
8	Case Studies of Convergence of Machine Learning and Laser Technology	68
9	Data-Driven Approaches in Laser Technology	83
10	Future Trends in Laser Technology	97
11	Do We Need Coding for Machine Learning in Lasers	108
12	Challenges, Opportunities, Conclusion, and Outlook	114
	References and Further Reading	121
	Appendix	126
	Acknowledgment	132
	Index	133

PREFACE

Laser technology has transformed our world in ways once only imagined. Lasers have become integral to modern society, from enabling life-saving surgeries to enhancing manufacturing speed and improving global communications. With the rise of machine learning and artificial intelligence, lasers are becoming smarter, more efficient, and more adaptable. This book, Empowering Laser Technology with Machine Learning, is a comprehensive guide designed to bridge the gap between the established field of laser technology and the rapidly advancing world of machine learning. It introduces readers to the exciting opportunities when these two fields converge, creating smarter, more adaptive, and more efficient laser systems capable of revolutionizing industries. Written with a broad audience in mind, this book caters to both seasoned professionals and newcomers. For those with foundational knowledge in laser technology, machine learning, or data processing, this book provides insights into integrating these technologies. Meanwhile, young researchers and curious readers will find that the book's clear language and straightforward explanations make even complex ideas accessible. One of the standout features of this book is its emphasis on using block diagrams and flowcharts. These diagrams help readers not only to understand complex topics but also to see how these principles can be implemented. Block diagrams and flowcharts guide readers through various real-world applications throughout the book. By using these, the book highlights essential processes in laser-machine learning integration, from predictive maintenance and adaptive beam shaping to laser-driven image recognition and material processing. Each diagram provides a step-by-step representation of processes, showing readers exactly how these systems operate and interact. Furthermore, the book is rich with case studies that provide in-depth examples of how laser technology and machine learning work together to solve real-world problems. Each case study is supported by diagrams, allowing readers to see the structure and flow of the entire system. Ultimately, Empowering Laser Technology with Machine Learning serves as both an educational resource and a practical toolkit. Whether you are a researcher looking to explore new frontiers, a student eager to learn the basics, or an industry professional seeking to apply machine learning to laser systems, this book provides a clear, structured path forward. The combination of explanations, real-world applications, block diagrams, and flowcharts makes this book an essential guide for anyone interested in understanding and harnessing the power of intelligent laser technology. This journey begins with an **Introduction** to set the stage, followed by an **Overview of Laser Technology**, which explores how lasers work and the fundamental principles driving their wide-ranging applications. **The Journey of Machine Learning** introduces readers to the evolution of machine learning, tracing its rise from academic curiosity to a pivotal technology reshaping industries worldwide. In **Artificial Intelligence and Machine Learning Hand in Hand**, we unpack the relationship between AI and machine learning, focusing on how these two areas complement each other. This sets the stage for the **Basics of Machine Learning**, where readers gain an understanding of essential concepts like supervised, unsupervised, and reinforcement learning, all explained with clarity and real-world relevance. The **Intersection of Laser Technology and Machine Learning brings** these fields together, revealing how machine learning can elevate laser performance, precision, and adaptability. In **Machine Learning Techniques in Laser Applications**, we look at real-life examples where machine learning enhances laser applications across various fields, including material processing, scientific research, and medical treatments. To reinforce understanding, we provide case studies supported by diagrams illustrating workflows, making connecting theory with practice easier. This chapter is particularly valuable for those interested in practical applications and implementation. The book then dives into **Advanced Laser Technology**, spotlighting cutting-edge developments and innovations pushing the boundaries of what lasers can achieve. **Case Studies of the Convergence of Machine Learning and Laser Technology** further explore in-depth examples where machine learning has directly improved laser applications, from ultrafast optics to real-time monitoring in high-precision procedures, backed by detailed block diagrams and flowcharts. These case studies provide a hands-on approach to understanding complex processes. Special attention is given to integrating Ti-sapphire-based Chirped Pulse Amplification (CPA) systems with machine learning, showcasing how optimization techniques can improve CPA operation, stability, and efficiency. This intersection between CPA and machine learning enables powerful advancements in high-power laser systems, making them more adaptable and effective for day-to-day use in labs and industries. **Data-Driven Approaches in Laser Technology** introduce data's role in making lasers more intelligent and reliable, illustrating how data-driven insights can improve laser design, functionality, and performance. **Future Trends in Laser Technology** contemplates the possibilities, exploring advancements and challenges in integrating machine learning with lasers, focusing on automated optimization, real-time adjustments, and enhanced diagnostic capabilities that ensure precise operation. Finally, **Do We Need Coding for Machine Learning Applications in Lasers**? delves into the balance between traditional coding and modern no-code platforms in applying machine learning to laser technology. It examines how these approaches can empower professionals to tackle challenges like beam quality optimization, predictive maintenance, and complex system integration while highlighting the hybrid methods that combine accessibility and customization. The book concludes with **Challenges, Opportunities, Conclusion, and Outlook**, where we summarize key takeaways and encourage readers to envision a future of laser technology that is adaptive, precise, and groundbreaking but also inclusive of innovative methodologies that make advanced laser applications more accessible to a wider audience.

Tayyab Imran
Magurele, Romania
November, 2024

Introduction

The interrelationship between scientific and technological innovations has never been more evident in the modern world, particularly in fields like photonics and artificial intelligence (AI). Among the most transformative inventions of the 20th century, the laser stands as a technological marvel with far-reaching applications in fields ranging from telecommunications to medical, manufacturing, and scientific research. Since its inception in the 1960s, laser technology has experienced remarkable advancements, from creating the first ruby laser to developing ultrafast lasers capable of generating femtosecond pulses, revolutionizing many sectors. On the other hand, machine learning (ML), a subset of AI, has witnessed a surge in interest and applications in recent decades, evolving from simple computational models to highly sophisticated algorithms capable of analyzing massive datasets and making autonomous decisions. Machine learning's ability to uncover patterns and make predictions from data without explicit programming has spurred innovation across multiple industries, including medical, finance, and cybersecurity. Its applications now stretch even into fields deeply rooted in physics and engineering, such as laser development, offering novel system optimization, fault detection, and predictive modeling approaches. The convergence of laser technology and machine learning represents a fascinating frontier where traditional photonics and modern computational intelligence meet. Lasers are inherently complex systems with numerous physical parameters, such as power, wavelength, intensity, pulse width, and beam quality, interacting in highly non-linear ways. Optimizing these parameters for different applications like medical lasers for precision surgery, industrial lasers for cutting and welding, or scientific lasers for experiments in ultrafast optics requires a deep understanding of the physics and underlying system dynamics. Historically, this optimization has relied heavily on empirical methods, time-consuming simulations, or expert manual tuning. However, as laser systems grow more sophisticated, with regenerative or multi-pass amplifiers, non-linear optical interactions, and pulse shaping techniques becoming commonplace, traditional methods of optimization and analysis are reaching their limits, where machine learning becomes invaluable. Machine learning algorithms, particularly those based on deep learning and neural networks, have the potential to process the vast amounts of data generated by laser systems and discern relationships between variables that are too complex for conventional methods to capture. This enables not only optimization but also real-time adaptive control of laser systems. For example, in ultra-intense laser applications, such as those using Ti-sapphire crystals and fiber lasers, small variations in system parameters can lead to significant changes in output pulse characteristics. Machine learning techniques can predict these outcomes more efficiently than conventional models.

Moreover, machine learning-based models can assist in automating the alignment of optical components, which is often a tedious and error-prone process, especially in multi-pass or regenerative amplifier systems. This can dramatically reduce time spent in experimental setups and improve the reliability of laser operation, particularly in high-stakes environments such as large-scale scientific facilities or precision medical treatments. Beyond system optimization, machine learning also shows promise in areas like fault detection and predictive maintenance for lasers. By continuously monitoring system performance, machine learning models can detect subtle patterns that indicate early-stage component degradation or misalignment, allowing for preventative maintenance before catastrophic failure occurs. This approach is particularly valuable in huge laser research laboratories like ELI-NP (www.eli-np.ro) and industrial laser system settings, where downtime due to laser failure can be costly, and in research environments, maintaining system integrity is critical for successful experiments.

Furthermore, machine learning models can be trained to adapt laser parameters in real-time, responding dynamically to changing conditions or variations in material properties in fields like laser-based manufacturing or material processing. In this sense, machine learning allows for unprecedented flexibility and control, pushing the limits of what laser systems can achieve. Integrating machine learning into the design, operation, and optimization of laser systems marks a significant step forward in both fields, with implications across diverse applications, from optimizing ultra-intense laser pulses to improving material processing and even enabling real-time system diagnostics, the use of machine learning in photonics heralds a new era of smarter, more efficient laser technologies.

OVERVIEW OF LASER TECHNOLOGY

LASER, which stands for Light Amplification by Stimulated Emission of Radiation, is a cornerstone of modern photonics. At its core, a laser operates on the principle of stimulated emission, where an electron in a higher energy state returns to a lower energy state, emitting a photon in the process ($E_1 - E_2 = \Delta E$). This phenomenon is crucial because it allows for light amplification, creating a coherent beam that is highly monochromatic and directional.

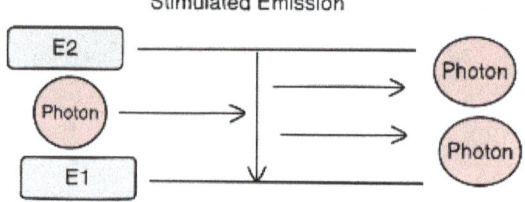

The essential components of a laser, illustrated in the block diagram, include a gain medium, an energy source (often referred to as a pump), and an optical cavity or resonator. Together, these elements facilitate the generation of laser light. The gain medium can be a solid-state, semiconductor, liquid, or gas, and its properties significantly influence the laser's output characteristics. Different materials can produce different wavelengths of light, which is why there is a wide variety of lasers, including semiconductor, dye, and fiber lasers. Each type has its unique advantages and applications. For instance, semiconductor lasers are compact and energy-efficient, making them ideal for consumer electronics, whereas solid-state and fiber lasers are known for their high power and efficiency, making them suitable for industrial cutting and welding applications. In addition to the gain medium, the pumping mechanism provides the necessary energy to excite the electrons. This energy can be supplied through electrical

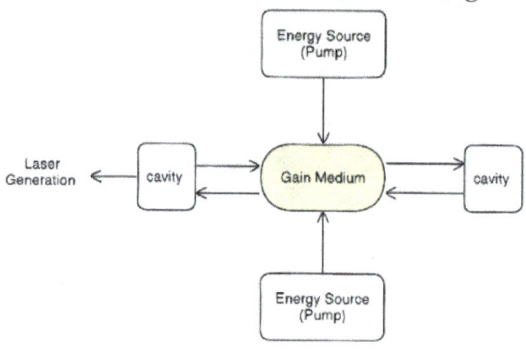

current, another light source (laser or flash lamp), or chemical reactions. The choice of pump source affects the efficiency of the laser and its overall performance. Understanding these fundamentals allows us to optimize laser designs for specific applications, leading to advancements in fields such as laser applications research, manufacturing, medical, and telecommunications.

Laser beams possess unique properties that differentiate them from other light sources; coherence, monochromatic, and directional contribute to their effectiveness in various applications. Coherence refers to the consistent phase relationship between the waves in the beam, and monochromaticity indicates that the light emitted is of a single wavelength; directionality ensures the beam can travel long distances with minimal divergence.

Monochromatic and Coherent

As machine learning continues to evolve, its integration with laser technology is becoming increasingly prevalent. Machine learning algorithms can optimize laser operation by analyzing data from various sources, enhancing efficiency, and improving real-time performance. This synergy opens new avenues for research and development, allowing for the creation of smarter laser systems capable of adapting to changing conditions and requirements. Leveraging machine learning, unlocking novel applications, and improving existing technologies pave the way for the future of laser innovation.

The block diagram offers a comprehensive overview of the fundamentals of laser technology and its integration with machine learning. It highlights in a simple way how machine learning can optimize laser operations, improving performance and enabling adaptive laser systems. The laser's core components, the energy source (pump), gain medium, and optical cavity collectively influence the laser's key properties: coherence (phase relation between waves), monochromaticity (single wavelength), and directionality (minimal divergence). These properties determine the laser's performance in applications.

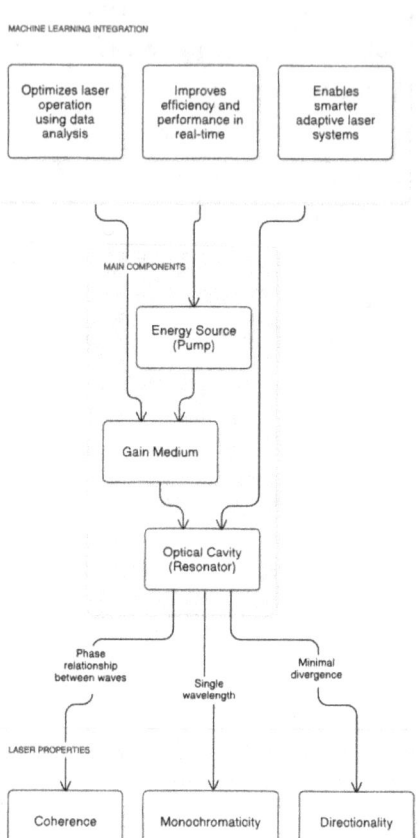

Overview of Laser Fundamentals

Historical Development of Laser Technology

The historical development of laser technology can be traced back to the early 20th century, with foundational theories in quantum mechanics and light behavior. The concept of stimulated emission, a crucial principle behind laser operation, was proposed by Albert Einstein in 1917. This theoretical groundwork set the stage for future advancements in light amplification. However, it wasn't until the mid-20th century that the first practical laser was realized, marking a significant milestone in physics and engineering. The invention of the maser in 1953 by Charles Townes and Arthur Leonard Schawlow, which amplified microwaves, paved the way for the creation of the laser, culminating in the first working laser prototype in 1960 by Theodore Maiman.

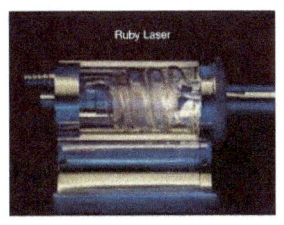

The 1960s saw a rapid expansion of laser technology as various types of lasers were developed, including gas, solid-state, and dye lasers. Each type brought unique properties and applications, further fueling research and innovation. The helium-neon laser, invented shortly after Maiman's ruby laser, became a staple in scientific research and industry due to its continuous wave output and ease of use. This era also witnessed the emergence of semiconductor lasers, which would eventually revolutionize telecommunications and consumer electronics, leading to the development of compact disc players and optical fiber communication systems. As the technology matured, the 1970s and 1980s introduced more sophisticated laser systems, including the advent of high-powered lasers used in industrial applications. These lasers were instrumental in material processing, cutting, and welding, enabling precision manufacturing techniques that were previously unattainable.

Additionally, the medical field began to explore laser applications, leading to breakthroughs in surgical procedures and treatments. The ability to high-precise light opened new avenues in medical science and engineering. The introduction of advanced laser systems in the 1990s and 2000s further diversified the field, with developments such as fiber and solid-state lasers becoming prominent. These innovations improved efficiency, reliability, and performance across various

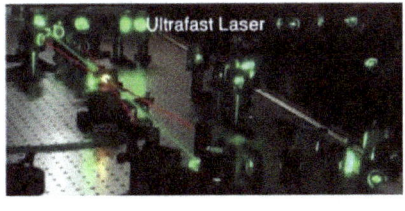

applications. Fiber lasers, in particular, offered significant advantages in beam quality and ease of integration into existing systems, making them popular in manufacturing and telecommunications. Concurrently, research into quantum and nonlinear optics contributed to understanding laser interactions and developing novel technologies like chirped pulse amplification (CPA), ultrafast lasers, and terahertz sources.

These days, the convergence of laser technology and machine learning presents exciting opportunities for future advancements. Increasingly leveraging machine learning algorithms to optimize laser performance, improve diagnostic capabilities, and enhance automation in laser systems. As we look forward, the potential for integrating artificial intelligence with laser technology promises to drive innovation in numerous fields, including materials science, medical, fundamental laser application research, and telecommunications. The historical development of laser technology highlights its transformative impact and sets the stage for a future where machine learning will play a pivotal role in shaping its trajectory.

This graph represents the timeline of major laser technology milestones, from Einstein's concept of stimulated emission in 1917 to AI-driven adaptive lasers in 2023. Each point on the graph marks a key development in laser technology over the years.

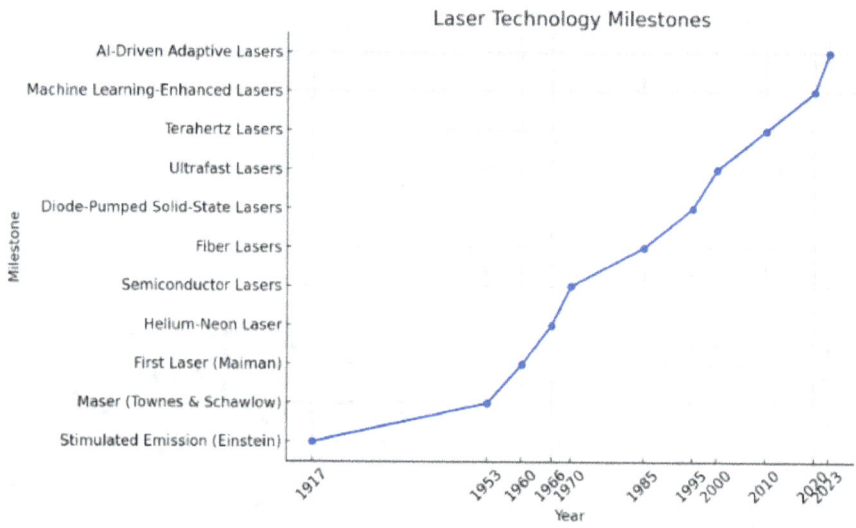

Applications of Lasers

Lasers are now a key part of many industries, proving their usefulness and effectiveness in various applications. Some important applications are? In the medical sector, lasers are employed for procedures such as laser surgery, phototherapy, and laser-assisted drug delivery. For instance, in ophthalmology, lasers are used for corrective eye surgeries like LASIK, where precision is paramount. The ability of lasers to focus energy on a specific target while minimizing damage to surrounding tissues has revolutionized surgical techniques. Additionally, lasers play a crucial role in diagnostic procedures 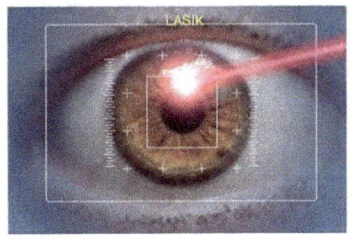 through techniques like laser-induced fluorescence, enabling healthcare professionals to detect diseases at early stages.

In communications, laser technology underpins the backbone of modern data transmission systems. Fiber-optic communication utilizes lasers to transmit information over long distances with minimal signal loss. The high bandwidth and speed of laser-based systems allow for rapid data transfer, which is essential for the functioning of the Internet and telecommunications. Furthermore, advancements in machine learning algorithms are enhancing the efficiency of these systems by optimizing signal processing and error correction, ensuring more reliable communication networks.

Manufacturing industries also benefit significantly from laser technology. Lasers are utilized in cutting, welding, engraving, and marking materials with high  precision and speed. Automating these processes, often enhanced by machine learning techniques, increases productivity and reduces waste. For instance, in automotive manufacturing, laser-cutting machines can achieve intricate designs while maintaining the structural integrity of materials. Machine learning algorithms improve these applications by enabling predictive maintenance, reducing downtime, and enhancing overall operational efficiency.

In research and development, lasers are powerful material analysis and characterization tools. Techniques like Raman spectroscopy and laser ablation provide insights into materials' chemical composition and properties. Machine learning is increasingly applied to interpret the vast amounts of data generated through these laser-based techniques, facilitating the identification of patterns and correlations that may not be apparent through traditional analysis methods. This synergy between lasers and machine learning is paving the way for breakthroughs in material science, nanotechnology, and other cutting-edge fields.

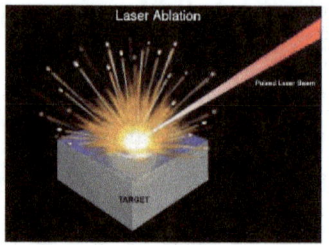

The entertainment industry has embraced laser technology for innovative applications, including light shows, projection mapping, and laser displays. These applications leverage the unique properties of lasers to create visually stunning effects that captivate audiences. With the integration of machine learning, the choreography of laser displays can be automated and optimized, creating dynamic and responsive performances that adapt to music and environmental conditions. As technology advances, the creative potential of lasers in entertainment will likely expand.

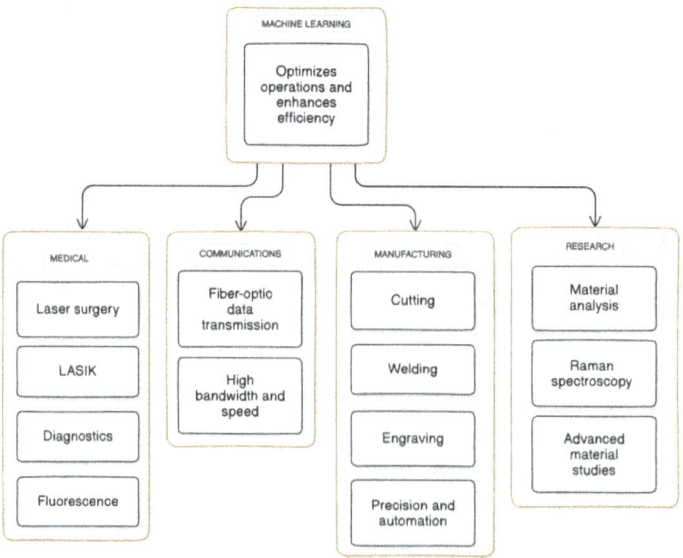

The block diagram illustrates the applications of laser technology, which have been discussed in previous paragraphs, across four major sectors:

medical, communications, manufacturing, and research. Machine learning is central to optimizing operations and improving efficiency in all these areas. In medical, lasers are used for surgeries (like LASIK), diagnostics, and fluorescence imaging. In communications, lasers facilitate high-speed fiber-optic data transmission. Manufacturing applications include cutting, welding, and precision automation. Research areas benefit from laser-driven techniques such as Raman spectroscopy and advanced material analysis.

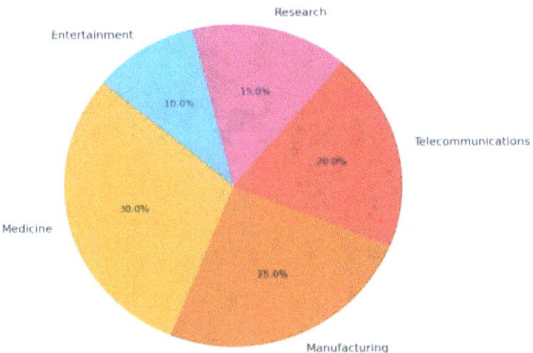

The applications of lasers their usage span various fields, each benefiting from their unique properties. The pie chart above provides a snapshot of the distribution of laser applications across different sectors based on available statistical data. It shows that lasers are most widely used in the medical field, accounting for 30% of applications, manufacturing at 25%, and telecommunications at 20%. Research and entertainment have smaller shares, at 15% and 10% respectively. This distribution highlights the versatility of laser technology in addressing needs from medical treatments to industrial processes and beyond.

THE JOURNEY OF MACHINE LEARNING

In 1950, British mathematician and computer scientist Alan Turing posed the key question, "Can machines think?" in his seminal paper, "Computing Machinery and Intelligence." Turing's exploration laid the groundwork for artificial intelligence (AI) and machine learning (ML) as we know them today. He introduced the concept of the Turing Test, a criterion for determining whether a machine can exhibit intelligent behavior indistinguishable from that of a human. This inquiry set the stage for subsequent investigations into the potential of machines to learn and adapt.

By the late 1950s and early 1960s, several foundational algorithms began to emerge, laying the groundwork for future advancements in machine learning. In 1957, Frank Rosenblatt developed the Perceptron, one of the first neural network models designed for pattern recognition. The Perceptron was a simple algorithm that could classify  input data into different categories based on learned weights, showcasing the potential for machines to identify patterns and make decisions. While it was limited in its capabilities, especially in its inability to solve problems that were not linearly separable, it generated enthusiasm and curiosity about the possibilities of machine learning. This initial excitement, however, was tampered with by the limitations of early models, which often failed to address the complexity of real-world tasks. The enthusiasm for machine learning peaked during the late 1950s and early 1960s and began to wane by the 1970s. This period is often called the "AI winter," characterized by a decline in funding and interest in AI research. The initial promise of machine learning clashed with the reality of its limitations. Many researchers discovered early algorithms could not handle complex tasks, leading to frustration and disillusionment. The funding for AI projects dried up, and many researchers switched to other fields. This stagnation lasted until the middle of the 1980s, when the landscape began to shift again, driven by advances in computer technology and a rebirth of interest in neural networks. In 1986, a pivotal moment occurred in machine learning when David Rumelhart, Geoffrey Hinton, and Ronald Williams published their influential paper on backpropagation. This algorithm allowed multi-layer neural networks to learn by efficiently calculating the gradient of the loss function. This breakthrough in training deep networks opened new avenues for research and applications. As the field evolved, various approaches to machine learning emerged, including decision trees, support vector machines, and ensemble methods. Each method brought unique advantages and was suited to different problems, further enriching the toolbox available to researchers. As the technological landscape evolved, the introduction of the internet in the 1990s provided an unprecedented abundance of data. This proliferation of information catalyzed machine learning research, allowing algorithms to be trained on larger datasets than ever before, thus enhancing their performance and accuracy. As the 21st century dawned, the era of big data transformed the machine-learning landscape. The exponential growth of data generated by social media, online transactions, and an increasing number of sensors created new opportunities for machine learning applications across various sectors. In 2006, Geoffrey Hinton and his team published a landmark paper on deep learning, a subset of machine learning that utilizes deep neural networks to model complex patterns in data. Deep learning algorithms can automatically extract features from raw data, making them particularly powerful for image and speech recognition tasks. This marked a significant shift in how machine learning could be applied, leading to groundbreaking technological advancements. Prominent technology companies began to harness the power of machine learning to enhance their services and improve user experiences. Google, for instance, implemented sophisticated machine learning algorithms to refine its search results, delivering more relevant information to users. Similarly, Netflix adopted machine learning to personalize content recommendations, significantly increasing user engagement and satisfaction. As these companies showcased the potential of machine learning, interest surged across various industries, prompting further research and investment in the field. Today, machine learning is not merely a niche area of study but a cornerstone of artificial intelligence, playing an integral role in diverse industries, including medical, finance, autonomous vehicles, laser technology, the electronics industry, and more. Techniques such as reinforcement learning have shown remarkable promise, enabling agents to learn optimal strategies through trial and error in complex environments. Meanwhile, advancements in hardware, particularly the development of graphics processing units (GPUs), have accelerated training times, allowing researchers to build more intricate models. As we look to the future, the potential of machine learning continues to expand exponentially, and the technical capabilities of machine learning are being explored. Machine learning has undergone significant transformations from its humble beginnings in the 1950s to its current status as a pivotal technology. As we continue to explore its capabilities and applications, machine learning will likely yield even more revolutionary changes in the years to come, further bridging the gap between human intelligence and machine capabilities.

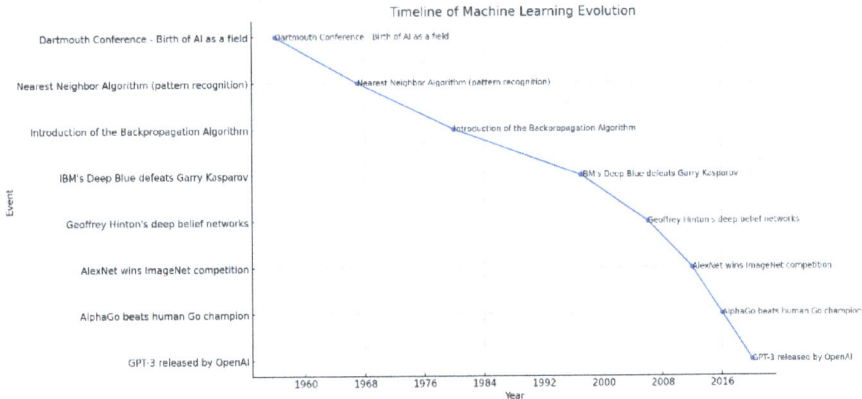

The timeline graph of the machine learning journey illustrates key milestones in the evolution of machine learning. Starting with Alan Turing's introduction of machine intelligence in 1950, it highlights significant developments such as the coining of machine learning in 1959, the introduction of deep learning in 2006, and the integration of machine learning across industries by 2020. It also shows periods of slow progress, known as AI winters, in the 1970s and 1980s.

Artificial Intelligence and Machine Learning Hand in Hand

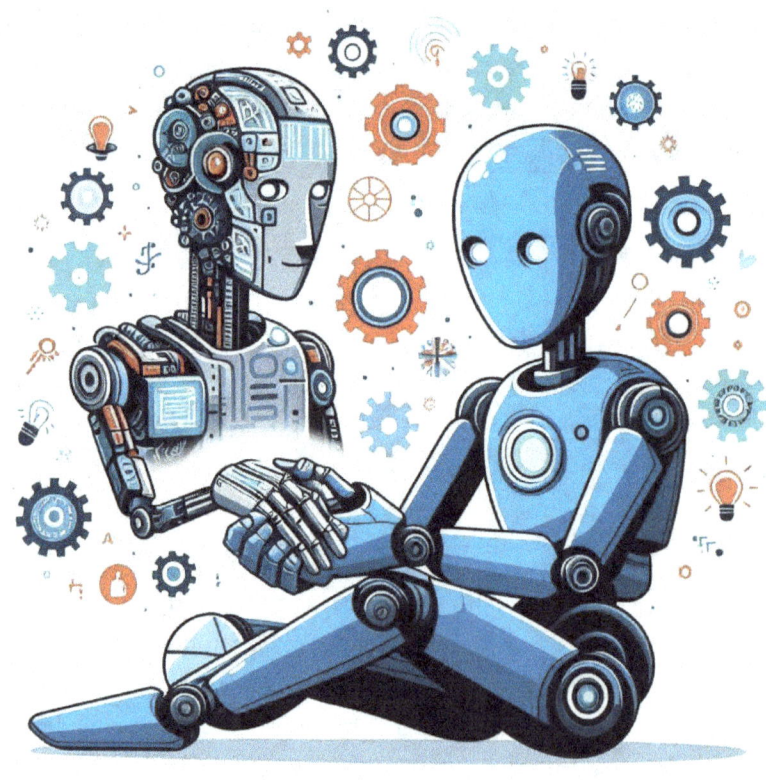

Artificial Intelligence (AI) and Machine Learning (ML) are among the most transformative technologies of our time, reshaping industries, economies, and daily life. While they are often used interchangeably, they have distinct meanings and applications. AI broadly encompasses the science of creating machines that exhibit intelligence similar to humans, capable of reasoning, learning, and making decisions. Machine learning, a specialized area within AI, focuses on developing systems that can learn autonomously from data patterns rather than relying on explicit programming. Together, AI and machine learning are driving unprecedented advancements in diverse fields, from medical and finance to laser technology and environmental science, continually pushing the boundaries of what technology can achieve. Understanding the distinct roles of AI and machine learning is essential to grasp their potential. AI represents the broad vision of creating machines with human-like intelligence and cognitive abilities, while machine learning provides the techniques that allow these systems to improve over time. AI is often divided into two major types: Narrow AI, which focuses on specific tasks, and General AI, which aims to replicate the full range of human cognitive abilities. While Narrow AI has become ubiquitous, General AI remains a long-term research goal. Narrow AI, seen in everyday life, powers virtual assistants like Siri and Alexa, which process voice commands and provide information, optimizes recommendation engines on platforms like Netflix and Amazon, and supports autonomous driving technologies by interpreting real-time sensor data to make driving decisions. These applications demonstrate how Narrow AI can significantly enhance functionality in specific domains. Machine learning is crucial to the development of adaptive, intelligent systems. Traditional programming requires explicitly defined rules, but machine learning allows systems to discover patterns and make predictions autonomously. Machine learning's core strength is its capacity to improve with more data, enabling applications like medical diagnostics, fraud detection, and natural language processing (NLP) to become more precise over time. Machine learning can be implemented using various approaches: supervised learning, where models learn from labeled data; unsupervised learning, which involves pattern detection in unlabeled data; and reinforcement learning, where models interact with an environment and learn through a reward system. These methodologies are foundational in applications such as medical imaging, customer segmentation, and autonomous vehicles, allowing AI to tackle increasingly complex challenges. One field where AI and machine learning are making notable contributions is laser technology, enhancing precision, adaptability, and efficiency. In manufacturing, AI and machine learning optimize laser settings for cutting, welding, and engraving tasks. By analyzing sensor data in real-time, these systems can dynamically adjust laser intensity, focus, and speed to suit specific materials or conditions. This leads to higher efficiency, less material waste, and improved product quality. For example, AI-controlled lasers can respond to changes in material density or thickness by automatically adjusting their parameters, ensuring consistent performance across various applications. In the medical sector, AI-driven laser systems are used in delicate procedures, such as laser eye surgery, where machine learning models fine-tune laser targeting to achieve safer, more precise outcomes. These AI-powered lasers allow for minimally invasive surgeries with greater accuracy and shorter recovery times. Moreover, AI and machine learning are critical in pushing the boundaries of laser-based research, particularly in fields like quantum computing, telecommunications, and environmental monitoring. In quantum computing, AI and machine learning algorithms manage laser systems that highly precisely control photons. These lasers are essential in stabilizing quantum experiments, where maintaining precise control over quantum particles is vital. In telecommunications, machine learning algorithms optimize laser modulation, facilitating high-speed data transmission in fiber-optic networks, which is crucial for advanced applications and future 6G technology. In environmental science, AI-powered laser systems are used for atmospheric monitoring, where lasers detect and measure gases like carbon dioxide and methane. Machine learning algorithms interpret the data here, helping scientists more accurately monitor pollution and greenhouse gases. This convergence of AI, machine learning, and laser technology opens new frontiers, enabling scientific breakthroughs and enhancing various applications from industrial processes to environmental sustainability. The future of AI and machine learning holds both exciting potential and significant responsibility. Quantum machine learning, an emerging field, aims to combine quantum computing's power with machine learning techniques to solve highly complex problems faster than classical computing allows. This combination could accelerate drug development, climate modeling, and materials science discoveries. If successful, quantum machine learning could revolutionize sectors by enabling AI systems to analyze massive datasets and derive insights at unprecedented speeds.

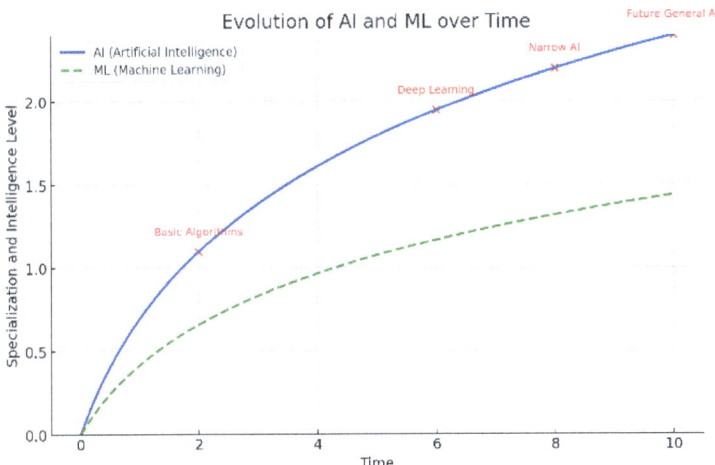

The 2D graph represents the progression of AI and machine learning over time, emphasizing how these fields have evolved in terms of intelligence and specialization. The blue line shows AI as a broad field advancing steadily, with increasing cognitive capabilities in applications over time. The green dashed line represents machine learning, a subset of AI, with its intelligence levels following a similar but more specialized trajectory, indicating its focus on data-driven learning and pattern recognition. Basic Algorithms, Deep Learning, and Narrow AI mark significant advancements in these fields. At the same time, Future General AI indicates ongoing research towards a system capable of a broader range of human-like cognitive functions.

Conclusively, AI and machine learning are interdependent, working to create intelligent systems that solve complex problems and improve quality of life. Machine learning enables AI's adaptability, allowing systems to refine themselves through experience and tackle sophisticated tasks. This partnership has sparked a technological revolution, reshaping industries, scientific research, and everyday experiences.

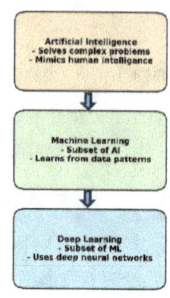

BASICS OF MACHINE LEARNING

Machine learning is a subset of artificial intelligence that focuses on developing algorithms that enable computers to learn from and make predictions or decisions based on data. It has gained significant attention recently due to its potential to transform various fields, including laser technology. Understanding the foundational concepts of machine learning is essential in the laser technology field, as it can enhance the efficiency and effectiveness of laser systems through predictive analytics and automation.

Introduction to Machine Learning Concepts

At its core, machine learning involves training models on datasets to recognize patterns and make informed decisions. This training process typically includes three main phases: data collection, model training, and model evaluation. In laser technology, extensive datasets comprising laser parameters, performance metrics, and environmental conditions are used to develop real-time models that optimize laser operations.

Leveraging these advanced techniques can improve operational efficiency, reduce costs, and enhance the overall performance of laser systems. There are several types of machine learning approaches, including supervised learning, unsupervised learning, and reinforcement learning. **Supervised learning** is characterized by using labeled datasets, where the model learns to predict outcomes based on input features. In laser technology applications, this could involve predicting the optimal settings for a laser system based on historical performance data. On the other hand, **unsupervised learning** does not rely on labeled data and focuses on identifying hidden patterns within the data. This approach can be particularly useful for anomaly detection in laser systems, where identifying unexpected behavior is crucial for maintaining system reliability. **Reinforcement** learning is another critical area in machine learning, where agents learn to make decisions by interacting with their environment and receiving feedback through rewards or penalties. In laser technology, reinforcement learning can be applied to optimize laser parameters dynamically during operation, adapting to changing conditions in real-time. This capability can lead to significant advancements in laser applications, such as materials processing and medical procedures, by ensuring optimal performance under varying circumstances.

The block diagram shows the integration of machine learning into laser technology, breaking it down into two main sections: core components and types of machine learning. The core components include data collection, model training, and model evaluation, which form the foundation for developing machine learning models. These models are then applied to laser technology applications through three machine learning approaches: reinforcement, supervised, and unsupervised. Each learning type optimizes laser performance by enabling smarter, more adaptive systems based on real-time data analysis and feedback.

Machine Learning and Laser Technology Integration

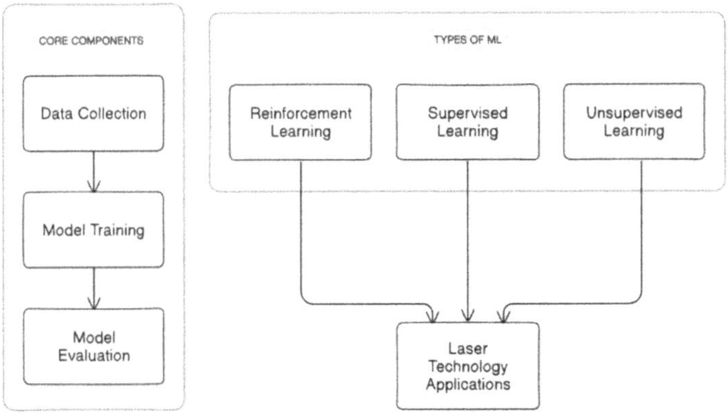

By understanding the fundamental principles of machine learning, the laser technology field can leverage these tools to drive advancements and improve system design.

Mathematical Foundations of Machine Learning

Machine learning is built on a foundation of mathematical principles that enable algorithms to analyze data, recognize patterns, and make predictions. These foundations include essential concepts in Linear Algebra, Probability and Statistics, and Calculus and Optimization.

Linear Algebra plays a central role in machine learning, particularly in the representation of data and the operations performed on it. Data in machine learning is often represented as matrices or vectors, and many algorithms rely on matrix operations such as multiplication, inversion, and eigenvalue decomposition. For instance, the weights of a neural network are stored in matrices, and matrix operations update these weights during training. One core operation is matrix multiplication, given by $C = A \times B$, where $C_{ij} = k\sum A_{ik} B_{kj}$.

Linear algebra is critical in modeling laser beam profiles and optimizing optical systems. For instance, in laser beam shaping, data about beam intensity across different spatial coordinates can be represented in a matrix. Multiplying this data with a transformation matrix allows for precise beam manipulation, such as focusing or expanding the beam for applications like laser cutting or welding.

Principal Component Analysis (PCA), a dimensionality reduction technique, relies on **eigenvectors** and **eigenvalues**. If A is a square matrix, the eigenvalue equation is $A\mathbf{v} = \lambda \mathbf{v}$, where λ is the eigenvalue and $\mathbf{v}$ is the eigenvector of matrix A. These eigenvectors help determine the principal components of data. PCA can be applied to analyze high-dimensional datasets from laser-based spectrometers, such as Raman or Brillouin spectroscopy systems. By identifying the principal components, PCA reduces noise and highlights the most significant features, enabling the detection of subtle material properties with high accuracy.

Probability and Statistics provide the framework for making predictions and quantifying uncertainty in machine learning models. Probability theory is foundational for many algorithms, especially those in unsupervised learning and probabilistic graphical models. Baye's Theorem, a key concept in probability, helps to update the probability of a hypothesis given new evidence $(A|B) = P(B|A)/P(B)$, where $P(A|B)$ is the posterior (the posterior distribution is central because it reflects the updated model's beliefs about a variable or parameter after considering observed data, Posterior = [(Likelihood×Prior)/Evidence] probability, $P(B|A)$ is the likelihood, $P(A)$ is the prior probability, and $P(B)$ is the evidence. Statistical methods are used to estimate the parameters of machine learning models, evaluate their performance, and validate predictions. For instance, the **Mean Squared Error (MSE)**, a common measure to evaluate model performance, is given by $MSE = \frac{1}{n}\sum_{i=1}^{n}(y_i - \hat{y}_i)^2$, where y_i are the actual values, $\hat{y}_i$ are the predicted values, and n is the number of observations. Probability and statistics are key to modeling uncertainty and making predictions in laser systems. For example, Bayes' Theorem can be used to optimize laser alignment in advanced optical setups. Given prior probabilities of alignment error and the likelihood of observing specific beam distortions, Bayes' Theorem calculates the posterior probability of misalignment. This guides precise adjustments, ensuring optimal beam quality and power; another example is assessing the performance of laser machining systems. Statistical methods such as Mean Squared Error (MSE) can evaluate the difference between the desired and actual cut profiles in a laser cutting or engraving application. Minimizing this error helps fine-tune the system for improved precision and consistency. **Calculus and Optimization** are crucial for optimizing machine learning models, particularly through **differential calculus**. Many algorithms aim to minimize a loss function, quantifying the difference between the model's predictions and the true values. The **gradient descent** algorithm, used to find the minimum of the loss function L(w) with respect to the weights w, is based on the gradient of L, $w = w - \eta \nabla L(w)$, where η is the learning rate, and $\nabla L(w)$ is the gradient of the loss function. The gradient represents the direction and rate of the fastest increase, and taking steps in the opposite direction minimizes the loss function. Optimization techniques driven by calculus are essential in laser technology to enhance system efficiency. Gradient descent, for instance, can be used to minimize a loss function representing a laser system's energy efficiency or output power. If parameters such as pump power, cooling rates, or cavity alignment affect this loss function, the gradient provides the direction for adjusting these parameters to improve performance. **Partial derivatives** are also essential, especially in complex functions with multiple variables, where the partial derivative of f(x,y) with respect to x is $\frac{\partial f}{\partial x} = \lim_{\Delta x \to 0} \frac{f(x+\Delta x, y) - f(x,y)}{\Delta x}$, Calculus-driven optimization methods like gradient descent enable iterative improvement of model parameters to minimize error. Partial derivatives are critical in the design and optimization of laser resonators. For example, the intensity distribution within a laser cavity depends on factors like mirror curvature and reflectivity. Iteratively refine the design to maximize output power and beam quality by calculating partial derivatives with respect to these variables. These mathematical foundations collectively empower machine learning algorithms to process data, uncover patterns, and make accurate predictions, facilitating a deeper understanding of complex datasets across various fields. When integrated with machine learning, these mathematical principles enable transformative advancements in laser technology.

Types of Machine Learning Algorithms

As discussed in the first part of this chapter, machine learning algorithms can be broadly categorized into three main types: supervised, unsupervised, and reinforcement. Each type serves distinct purposes and employs different methodologies, making them suitable for various applications within laser technology. Supervised learning involves training a model on labeled datasets, where the algorithm learns to map input data to the correct output. This approach is particularly useful in applications such as predicting the performance of laser systems or optimizing parameters for laser machining processes based on historical data. On the other hand, unsupervised learning deals with unlabeled data, enabling the algorithm to identify patterns and structures without explicit guidance. This type of learning can be invaluable in analyzing complex datasets generated from laser experiments, such as identifying clusters of similar outcomes in laser material interactions. Techniques such as clustering (grouping similar data) and dimensionality reduction (simplifying complex data) can reveal hidden patterns, which may inspire innovations in the development and application of laser technology. Reinforcement learning presents a different approach, where an agent learns to make decisions by interacting with an environment. In laser technology, this can involve optimizing laser parameters in real-time for applications like adaptive optics or laser welding. The agent receives feedback from its actions and adjusts its strategy to maximize a reward signal, which can be defined in performance metrics such as the finished product's efficiency, precision, or quality.

Numerous specific algorithms enable researchers and engineers to address various challenges within these broad categories. For example, commonly used supervised learning algorithms include linear regression, support vector machines, and neural networks, each offering unique strengths and applications. **Linear regression** is often applied for predicting continuous outcomes based on input variables, with a simple linear model defined by $y = \beta_0 + \beta_1 x_1 + \beta_2 x_2 + ... + \beta_n x_n + \varepsilon$: y is the predicted outcome, β_0 is the intercept, β_n are the coefficients for each input variable x_n, and ε represents a random error. Linear regression is applied to predict outcomes like laser beam quality, energy deposition, or material processing results based on power, intensity, pulse duration (in case of pulsed laser) and wavelength parameters. Using a simple linear model makes it possible to fine-tune a laser system's operational parameters for applications such as laser cutting, additive manufacturing, or precision surgery. **Support vector machines (SVMs)** excel in classification tasks by finding optimal boundaries between data points and maximizing the margin between classes in a dataset. SVMs classify complex data by distinguishing between successful and suboptimal laser ablation patterns or categorizing material responses to laser pulses in processes like micromachining or laser annealing. **Neural networks**, particularly deep learning models, are powerful for handling complex patterns in large datasets and learning features hierarchically (something organized in a ranked or graded order) through layers of neurons. These models are instrumental in more advanced applications, including designing adaptive optics for laser beam shaping, predicting thermal effects in laser-tissue interactions, and optimizing ultrafast laser pulse compression for enhanced imaging or material processing precision.

In laser technology, these algorithms can predict outcomes based on input variables such as laser power, energy, beam profiling, intensity, and material properties, enabling precise control and optimization. Meanwhile, algorithms like k-means clustering and principal component analysis (PCA) are commonly used for unsupervised learning, uncovering hidden patterns and trends without labeled data. **K-means** clustering aims to partition data into k clusters, where each data point belongs to the cluster with the nearest mean. This process minimizes the sum of squared distances between points and their cluster centers; mathematical expression can be $\min \sum_{i=1}^{k} \sum_{x \in C_i} \|x - \mu_i\|^2$: C_i is a cluster, and μ_i is its mean. K-means clustering is valuable for segmenting data related to laser-material interactions, such as grouping materials with similar absorption characteristics or thermal thresholds. This is crucial in applications such as developing coatings for laser mirrors or selecting materials for specific wavelengths in photonics. **Principal component analysis (PCA)** reduces the dimensionality of data by transforming variables into a set of uncorrelated principal components. These components capture the maximum variance, allowing for easier visualization and understanding of data trends. The transformation is achieved by solving for eigenvectors and eigenvalues of the covariance matrix Σ: $\Sigma \upsilon = \lambda \upsilon$, where υ represents the eigenvectors (principal components) and λ are the eigenvalues. PCA simplifies large and complex datasets, such as spectral data from ultrafast lasers, enabling researchers to identify key factors influencing pulse duration, energy distribution, or spectral broadening, essential for optimizing systems used in medical imaging or advanced spectroscopy.

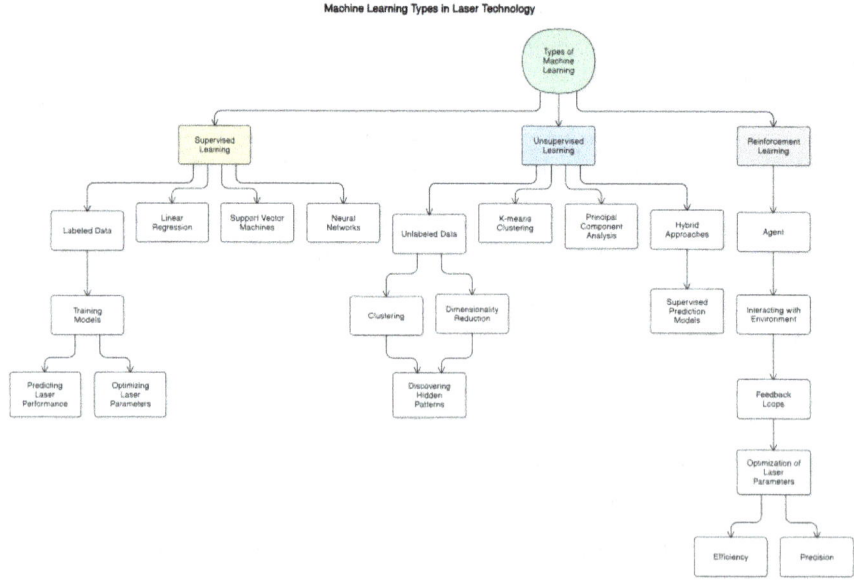

This organization of machine learning types is illustrated systematically in the accompanying block diagram, clarifying how each method fits within the broader artificial intelligence framework, supervised learning with algorithms like linear regression, support vector machines, and neural networks, which predict laser performance and optimize parameters based on labeled data. These models are trained to accurately predict specific laser attributes, such as power and material properties.

On the other side, unsupervised learning includes k-means clustering and principal component analysis, which work with unlabeled data to uncover hidden patterns or reduce dimensionality, helping optimize laser technology processes. Additionally, reinforcement learning, depicted with an agent interacting with the environment, uses feedback loops to optimize laser settings, improving efficiency and precision over time.

Hybrid Approaches in Machine Learning for Laser Technology

```
                    ┌─────────────┐
                    │   Hybrid    │
                    │Approaches in│
                    │  Machine    │
                    │  Learning   │
                    └──────┬──────┘
                  ┌────────┴────────┐
                  ▼                 ▼
           ┌─────────────┐   ┌─────────────┐
           │Unsupervised │   │ Supervised  │
           │  Learning   │   │  Learning   │
           └──────┬──────┘   └──────┬──────┘
                  ▼                 ▼
           ┌─────────────┐   ┌─────────────┐
           │Preprocessing│   │ Prediction  │
           │   Pattern   │   │Optimization │
           │  Discovery  │   │             │
           └─────────────┘   └─────────────┘
```

Hybrid approaches combining elements of different algorithms are also gaining traction. For example, integrating supervised and unsupervised methods can lead to more robust models that leverage the strengths of both techniques. In laser technology, this could mean using unsupervised learning to preprocess data before applying supervised models for prediction or optimization.

Key Terminology in Machine Learning

Understanding key terminology in machine learning is essential for anyone engaged in the intersection of laser technology and machine learning. Machine learning, a subset of artificial intelligence, involves algorithms that allow computers to learn from and make predictions based on data. Familiarity with foundational terms such as supervised, unsupervised, and reinforcement learning is crucial. As discussed, supervised learning entails training a model on a labeled dataset, where the desired output is known, while unsupervised learning deals with unlabeled data, enabling the model to identify patterns and structures within the data. Reinforcement learning, on the other hand, focuses on training agents to make decisions through trial and error, receiving feedback from their actions.

Another critical term is **feature**, which refers to an individual's measurable property or characteristic of an observed phenomenon. In the context of laser technology, features might include power, wavelength, pulse duration, or intensity of the laser beam. The selection and engineering of features significantly impact the performance of machine learning models. Effective feature extraction can improve accuracy in predicting outcomes, such as optimizing the parameters for laser cutting or enhancing imaging capabilities in laser diagnostics.

Model is another fundamental concept that represents the mathematical representation of a process or system. In machine learning, models are trained on data to capture underlying patterns, which can then be applied to make predictions or decisions. We discussed different models, such as decision trees, neural networks, and support vector machines, that can be utilized depending on the problem's complexity and the data's nature. In laser technology, models can be employed to predict how different laser parameters affect material properties or to enhance the automation of laser-based processes.

Training and testing are essential to developing machine learning applications. The training phase involves feeding the model a large dataset to learn from, while the testing phase evaluates the model's performance on unseen data to assess its generalization ability. Proper training and testing methodologies are vital to ensure that the model does not overfit the training data, which can lead to poor performance in real-world applications. This is particularly important in laser technology, where precision and reliability are important. **Algorithm** refers to the procedure or formula used to solve problems and make predictions in machine learning. Various algorithms optimize model performance, such as gradient descent, random forests, and convolutional neural networks (CNNs), play vital roles in optimizing model performance, each suited to different tasks. **Gradient descent** minimizes the error between predicted and actual values by iteratively adjusting model parameters toward the steepest descent. **Random forests**, an ensemble method, build multiple decision trees and average their predictions to improve accuracy, making them effective for classification and regression. **Convolution neural networks (CNNs)**, designed for image analysis, use convolutional layers to extract hierarchical features, allowing for precise pattern recognition in visual data. The choice of algorithm significantly influences outcomes in applications like laser technology, where models are used to predict and optimize parameters, assess material quality, and enhance precision in processing. By understanding these key terminologies, a learner can better navigate the evolving landscape where machine learning and laser technology converge, ultimately driving innovation and advancements in the field. The block diagram depicts and illustrates the above description of key terminology in machine learning.

Key Terminology in Machine Learning

It shows the relationships between feature extraction, algorithms, and learning types. It illustrates how different algorithms (e.g., gradient descent, random forests, CNNs) and data features (e.g., wavelength, power) apply to supervised, unsupervised, and reinforcement learning approaches, with model training and testing involved across all categories.

INTERSECTION OF LASER TECHNOLOGY AND MACHINE LEARNING

Machine learning has become a cornerstone in advancing laser technology, offering solutions for optimizing system design, enhancing manufacturing processes, and enabling sophisticated real-time control mechanisms. Traditional methods of laser development rely heavily on trial-and-error approaches and empirical rules, which are often limited by human intuition and can be inefficient. Machine learning leverages large datasets and computational power to model complex relationships within laser systems, enabling data-driven optimizations that enhance system performance, precision, and adaptability.

The Role of Machine Learning in Laser Development

In the design phase, machine learning algorithms are instrumental in exploring the high-dimensional parameter space of laser systems. Key parameters, such as the choice of the gain medium, cavity geometry, pumping configuration, and output couplers, significantly impact the laser's overall performance, including its efficiency, coherence, stability, and beam quality. Using historical and experimental data, machine learning models, particularly regression algorithms and neural networks, predict performance metrics based on these parameters. By training on large datasets, machine learning algorithms can identify optimal configurations that maximize desired characteristics while minimizing drawbacks, such as noise and thermal effects. Multi-objective optimization techniques, such as genetic algorithms combined with machine learning, allow the exploration of trade-offs between competing objectives, such as beam quality, efficiency, and cost. This combined approach enables a more exhaustive exploration of design possibilities, potentially leading to novel configurations with improved performance. Machine learning is critical in manufacturing in quality assurance, predictive maintenance, and process optimization. For example, predictive maintenance models employ supervised learning to analyze operational data from manufacturing equipment, identifying patterns that precede equipment failure or degradation in performance. These models use techniques such as anomaly detection, where deviations from normal operation patterns indicate potential faults. By forecasting maintenance needs, machine learning helps reduce unplanned downtime and extends the operational life of critical components like optics, pump sources, and cooling systems. Process optimization techniques leverage reinforcement learning to fine-tune complex production steps, such as the precise alignment of optical components or the deposition of coatings. Reinforcement learning agents learn optimal actions by interacting with the manufacturing environment, minimizing alignment errors or defects in coatings to achieve higher yields and tighter tolerance levels.

Machine learning enhances the laser system's ability to adjust dynamically to environmental changes or operational demands in real-time monitoring and adaptive control. Adaptive optics, for example, corrects wavefront distortions caused by atmospheric turbulence or optical aberrations in real-time. Machine learning models, particularly convolutional neural networks (CNNs) and support vector machines (SVMs), analyze sensor data from wavefront detectors, calculating adjustments to deformable mirrors or spatial light modulators. This process maintains high beam quality in free-space optical communication or high-resolution imaging, where aberrations can degrade performance. For closed-loop control, machine learning-based feedback systems use deep reinforcement learning to regulate parameters like laser energy per pulse, pulse duration, and repetition rate (in the case of pulsed lasers). In deep reinforcement learning, the algorithm learns a policy to maximize performance objectives, such as output stability or minimized pulse jitter, by adjusting laser settings based on real-time feedback. This is particularly advantageous in ultrafast laser applications where precise control over pulse characteristics is essential for high-speed, high-precision applications, such as micromachining or medical diagnostics.

Integrating machine learning across all stages of laser development, from parameter selection in design to adaptive control in operation, can significantly enhance system efficiency, reduce production costs, and push the boundaries of laser performance. The application of machine learning in laser technology thus enables a level of optimization and adaptability that would be challenging, if not impossible, to achieve using traditional methods alone.

Synergies Between Laser Applications and Machine Learning

Machine learning transforms laser technology by optimizing existing applications and enabling new, advanced functionalities across various fields. Integrating machine learning techniques with laser systems allows for dynamic parameter tuning, predictive analytics, and real-time adjustments that enhance precision, efficiency, and adaptability. In laser-based manufacturing, especially in additive manufacturing processes like selective laser sintering and laser melting, machine learning algorithms are employed to optimize critical parameters such as laser power, scan speed, and layer thickness. Traditional methods of setting these parameters often involve fixed values or manual adjustments, resulting in variability in the quality of 3D-printed components. By analyzing data from previous prints, machine learning models can predict how parameter changes will impact material density, surface finish, and structural integrity. This predictive capability enables manufacturers to achieve greater control over the process, ensuring consistent quality and reducing the need for post-processing or rework. Furthermore, machine learning algorithms, such as reinforcement learning, can adaptively adjust printing parameters on the fly in response to real-time data, enhancing precision and throughput. In the biomedical field, machine learning combined with laser technology has led to highly personalized treatment options, particularly in laser-based medical procedures. For instance, laser ablation, a method used for tissue removal in surgical settings, can be significantly enhanced with the help of machine-learning algorithms that analyze patient-specific data. Machine learning models can determine optimal laser settings, such as center wavelength, intensity, pulse, and power level, by processing information on tissue type, density, and patient health parameters for each unique procedure. This approach minimizes collateral tissue damage, reduces recovery times, and increases the success rate of surgeries. In photothermal therapies and other laser-based treatments for conditions like cancer, machine learning algorithms are used to predict tissue heat diffusion patterns, helping control laser application precisely to maximize therapeutic effects while minimizing side effects. Machine learning's predictive capabilities also allow for adjustments during the procedure, adapting to any unexpected tissue responses in real-time, thus enhancing the efficacy and safety of laser-based therapies. Machine learning has also shown significant promise in improving laser communication systems, particularly in free-space optical (FSO) communication. FSO communication systems, which use laser beams to transmit data over open spaces, are highly susceptible to environmental factors like atmospheric turbulence, temperature fluctuations, and dust or fog interference. These factors introduce noise and data loss, which can degrade communication quality, especially over long distances. Machine learning models, such as supervised learning algorithms and neural networks, analyze environmental data to predict atmospheric conditions and adjust the laser modulation techniques accordingly. Machine learning algorithms reduce transmission errors and enhance data rates by optimizing parameters like beam intensity, modulation format, and error correction protocols. Reinforcement learning techniques can improve FSO systems by enabling them to adjust dynamically to real-time environmental changes, maintaining a stable connection despite fluctuations. This adaptability is crucial for applications like satellite communications and drone-based laser links, where maintaining reliable data links over varying atmospheric conditions is essential for performance. Moreover, combining laser scanning technologies with machine learning revolutionizes autonomous navigation and robotics. Laser scanners can collect detailed spatial data, which machine learning algorithms can process to create accurate maps and models of environments. This capability is particularly valuable in applications ranging from autonomous vehicles to drone navigation. By synergizing these technologies, researchers can develop systems that perceive their surroundings more effectively and adapt to real-time dynamic conditions, enhancing their operational reliability and safety. Machine learning with laser diagnostics is paving the way for advanced research and development in various scientific fields. Lasers are often used in spectroscopy, imaging, and other diagnostic techniques, producing vast amounts of data that can be challenging to analyze manually. Machine learning techniques can automate data analysis, uncovering hidden correlations and insights that might go unnoticed. This synergy facilitates the exploration of new materials, the development of novel therapies in medical, and the advancement of fundamental scientific research, thus fostering innovation and discovery across multiple disciplines.

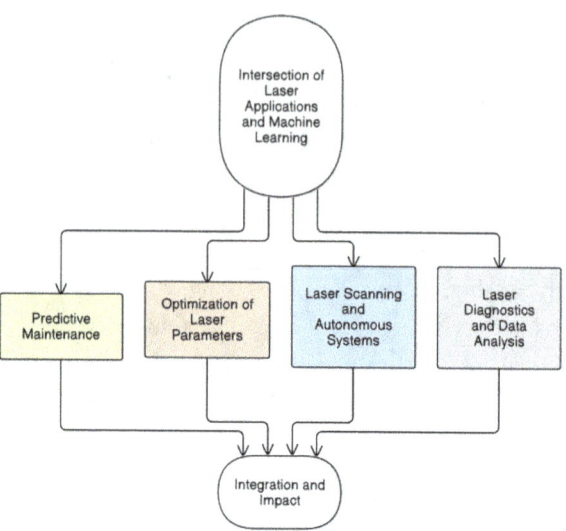

The block diagram illustrates the synergies between laser applications and machine learning, focusing on key areas where they intersect. These areas include predictive maintenance, optimization of laser parameters, laser scanning and autonomous systems, and laser diagnostics and data analysis, collectively contributing to enhanced integration and impactful outcomes in laser technologies.

MACHINE LEARNING TECHNIQUES IN LASER APPLICATIONS

Machine learning techniques in laser applications are revolutionizing fields like imaging, material processing, and optical communication by optimizing performance and enhancing precision.

Predictive Modeling in Laser Processes

Predictive Predictive modeling in laser processes leverages advanced algorithms and machine learning techniques to forecast outcomes based on historical data and real-time measurements. The complexity of laser interactions with different materials necessitates a robust modeling framework that can accommodate various parameters, such as laser power, intensity, beam profiling, pulse duration in case of a pulsed laser, and material properties. By utilizing machine learning models, we can analyze intricate patterns within large datasets, allowing them to predict the effects of specific laser settings on material behaviors. This capability enhances process efficiency and minimizes the traditional trial-and-error approach in laser applications. One of the primary advantages of predictive modeling in laser processes is its ability to optimize manufacturing techniques. For instance, in additive manufacturing, where lasers fuse materials layer by layer, predictive models can identify optimal parameters that improve material strength and reduce defects. Engineers can refine their processes before implementation by simulating different laser conditions and their impacts on the final product. This results in significant cost savings and waste reduction, as manufacturers can consistently achieve desired outcomes. The integration of predictive modeling also facilitates the exploration of new materials and laser applications. Machine learning algorithms can analyze vast datasets from various experiments, revealing insights that may not be immediately apparent through conventional analysis. This approach can uncover novel interactions between laser parameters and material responses, leading to innovative biomedical engineering and microfabrication applications. We can systematically investigate how different laser configurations affect new materials, paving the way for breakthroughs that expand the capabilities of laser technology.

Furthermore, predictive modeling enhances the understanding of laser-material interactions at a fundamental level. By employing regression analysis and neural networks, scientists can develop models that describe how energy absorption, thermal diffusion, and phase changes occur during laser processing. These models can serve as valuable tools, providing a framework for complex concepts related to laser physics and materials science. Visualizing and predicting outcomes can develop a deeper understanding of the intricacies involved in laser applications. Predictive modeling represents a transformative approach to enhancing laser processes through machine learning. By enabling more accurate predictions of outcomes based on various parameters, this technology improves efficiency and cost-effectiveness and fosters innovation in the field. As these models refine, the potential for discoveries and applications in laser technology will expand, ultimately shaping the future of manufacturing and material science.

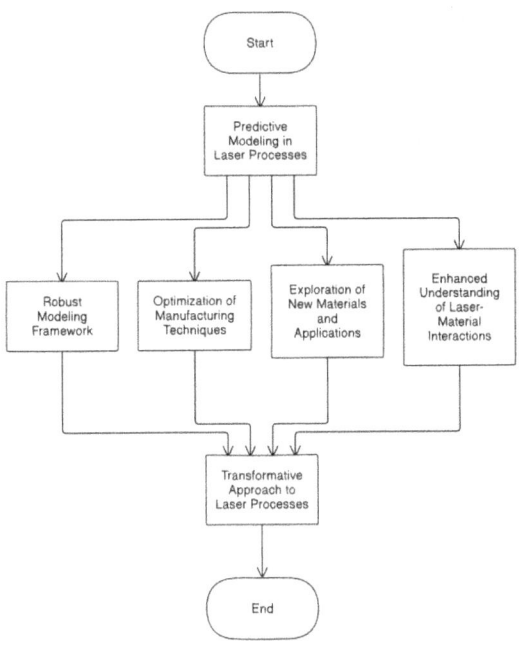

Block diagrammatically, starting from the predictive modeling stage; it shows how this approach branches out into four key areas: robust modeling framework, optimization of manufacturing techniques, exploration of new materials and applications, and enhanced understanding of laser-material interactions. These components then converge, leading to a transformative approach to laser processes. The final block, labeled end, signifies the culmination of these efforts in achieving advanced laser process methodologies. The diagram effectively shows how predictive modeling can drive innovation and optimization in laser technologies across various domains.

Image Processing and Computer Vision with Lasers

Image processing and computer vision are critical fields that significantly benefit from laser technology advancements. Lasers provide high-resolution, high-contrast imaging capabilities that enhance the ability to capture and analyze visual data. This is particularly relevant in 3D scanning, remote sensing, and biomedical imaging applications. We can achieve precise distances and surface characteristics measurements by utilizing lasers, improving the accuracy of image processing tasks. Integrating lasers with machine learning algorithms further enhances the capability to interpret and analyze the vast amounts of visual data generated. In the realm of computer vision, lasers enable the development of sophisticated systems that can recognize and understand complex environments. For instance, laser-based **LiDAR (Light Detection And Ranging)** systems are extensively used in autonomous vehicles for real-time mapping and obstacle detection. These systems generate detailed point clouds that machine learning models can process to identify objects and navigate safely. The high spatial resolution provided by laser systems allows for better differentiation between objects, reducing the likelihood of misclassification. This synergy between laser technology and machine learning revolutionizes how machines perceive their surroundings.

The diagram above illustrates the integration of laser and machine learning in computer vision systems, potentially using LiDAR technology as a foundational component, showcasing a step-by-step workflow. It begins with laser system integration, which enables laser-based imaging and data collection to gather structured visual and environmental data.

This data is then processed through laser data acquisition, which provides raw input for downstream tasks. The data undergoes machine learning model training for basic interpretation, equipping the system with foundational analytical capabilities.

Laser and Machine Learning Integrationicon in Computer Vision

In parallel, a high-spatial-resolution model for precision classification is trained to enhance detailed image analysis. Environmental and scene data recognition is achieved through computer vision for object detection and classification, leveraging machine learning to distinguish between objects in the scene. This information feeds into real-time mapping and obstacle detection, essential for navigation and interactive applications. Finally, all insights culminate in an integrated machine representation, enabling a cohesive understanding of the environment for advanced computer vision applications, with feedback loops refining the system's accuracy and adaptability. This synergy between laser technology and machine learning optimizes real-time decision-making in complex environments, such as autonomous navigation.

Additionally, lasers enhance imaging techniques in biomedical applications. In medical diagnostics, laser-based imaging modalities such as **Optical Coherence Tomography (OCT)** provide detailed cross-sectional images of tissues. These images are crucial for detecting early-stage diseases, particularly in ophthalmology and oncology. Machine learning algorithms can be applied to analyze these images, identifying patterns and anomalies that may not be immediately apparent to the human eye. This combination of lasers and machine learning is paving the way for more accurate diagnostics and personalized medical treatment.

This diagram represents the process of OCT in medicine, detailing a systematic approach to high-resolution medical imaging and diagnostic support. It begins by clarifying each step to ensure accuracy and consistency. OCT uses lasers to generate cross-sectional images of tissues, capturing fine structural details essential for accurate diagnostics. These detailed tissue images are then processed using machine learning algorithms to analyze and interpret the data, enhancing diagnostic accuracy. Combining OCT with other imaging modalities, such as MRI, provides a more comprehensive view, improving disease detection and assessment capabilities and insights from OCT imaging that support clinical decisions and treatment planning. The workflow includes periodic reviews to refine imaging techniques, implement improvements, and enhance imaging quality. Evaluations of treatment outcomes contribute to the continuous updating of algorithms with new data, ensuring that OCT technology evolves to meet medical needs more effectively. This feedback loop allows OCT to play a pivotal role in advancing precision medicine and improving patient care outcomes.

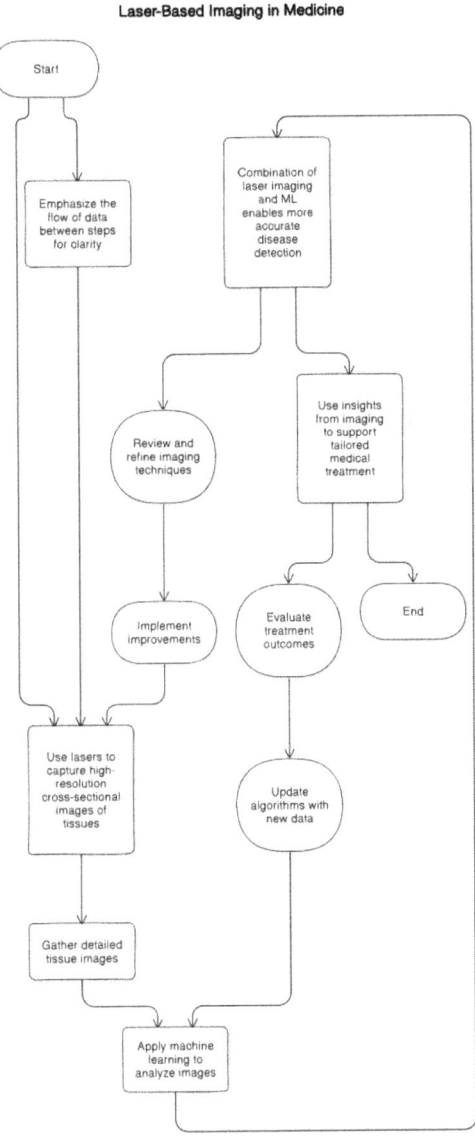

Laser-Based Imaging in Medicine

Another significant application of laser technology in **Image Processing is in the field of industrial automation**. Laser scanning techniques allow for accurately assessing components and assemblies in manufacturing processes. By integrating machine learning, systems can learn from historical data to predict failures or anomalies in the production line. This predictive maintenance approach improves operational efficiency and minimizes downtime and costs associated with unexpected equipment failures. Analyzing and processing images in real-time enhances the quality control processes in various industries.

Feedback Loop from Quality Control to Image Acquisition

```
                    Start
                      │
          ┌───────────┘
   IMAGE ACQUISITION
          │
    Capture Image
          │
    Preprocess
      Image
          │
   DATA ANALYSIS
          │
      Analyze          Request Re-
   Historical Data     acquisition
          │
   QUALITY CONTROL
          │
     Review Image
       ┌──┴──┐
    Meets   Does Not Meet
  Standards   Standards
      │         │
  Approve    Reject
   Image     Image
      │
     End
```

This diagram summarizes what is explained in the last paragraph and shows a quality control process with feedback for image processing in industrial automation. Images are captured, preprocessed, and analyzed against historical data to check if they meet quality standards. Approved images proceed, while rejected ones trigger a feedback loop for re-acquisition, ensuring consistent quality in production.

Automation and Control Systems

Automation and control systems are crucial in advancing laser technology, particularly when integrated with machine learning algorithms. Automation refers to using technology to perform tasks with minimal human intervention, while control systems are designed to manage and regulate the behavior of various devices and processes. In the context of laser technology, these systems enhance precision, efficiency, and adaptability across multiple applications, from manufacturing to medical. Incorporating machine learning into these systems allows for real-time data analysis and adaptive responses, making laser operations more intelligent and responsive to changing conditions.

One of the primary benefits of integrating machine learning with automation and control systems in laser technology is enhancing process optimization. Machine learning algorithms can analyze vast datasets generated during laser operations, identifying patterns that can lead to more efficient settings and improved outcomes. For instance, machine learning can optimize parameters such as power, intensity, and beam profiling and focus on laser cutting or welding processes, ensuring consistent quality and reducing waste.

Moreover, machine learning enhances predictive maintenance capabilities within laser systems. Traditional control systems often rely on scheduled maintenance, which may not align with the component's wear and tear. By employing machine learning techniques, these systems can monitor real-time performance metrics and predict when maintenance is needed, reducing downtime and extending the lifespan of the equipment.

In addition to optimization and maintenance, automation and control systems powered by machine learning can significantly improve safety in laser operations. Laser technology, while highly effective, poses potential hazards, especially in industrial settings. Advanced control systems can use machine learning to assess environmental variables and operator behaviors, implementing real-time safety protocols. For example, systems can automatically shut down or adjust laser parameters if they detect unsafe conditions, thus protecting personnel and equipment. Integrating safety features is essential for fostering a safety culture in environments where lasers are employed.

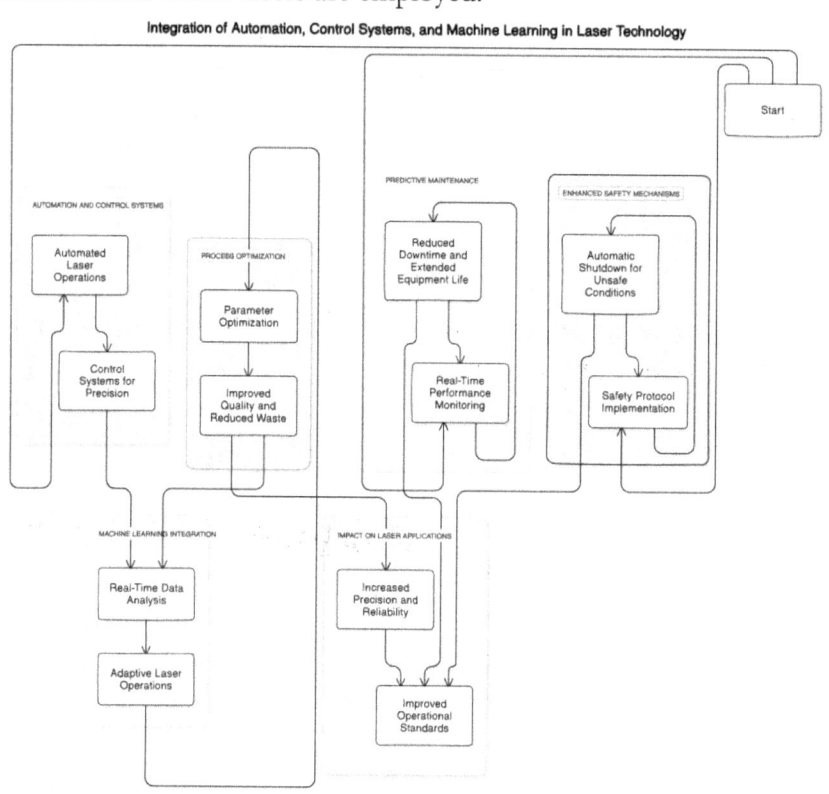

The diagram illustrates the automation, control systems, and machine learning integration in laser technology, detailing data flow, control processes, and feedback loops that enhance laser applications. It begins with automation and control systems, where automated operations and precision control systems set the foundation for efficient laser processes. Machine learning integration then enables real-time data analysis and adaptive responses, allowing the system to adjust laser parameters dynamically based on conditions. This data flow supports process optimization, where machine learning identifies optimal settings, improving quality and reducing waste. The predictive maintenance section shows how real-time performance monitoring enables predictive maintenance, reducing downtime and extending equipment life. Feedback loops ensure that data from performance monitoring continuously informs adaptive operations, making maintenance proactive. Enhanced safety mechanisms use machine learning to implement safety protocols that automatically adjust or shut down laser operations if unsafe conditions are detected, ensuring a safer operational environment. Finally, the impact on the laser applications section highlights the resulting benefits: increased precision, reliability, and higher operational standards. These elements demonstrate how automation, machine learning, and control systems collaboratively enhance efficiency, safety, and adaptability in laser applications.

The interplay between automation, control systems, and machine learning will shape its future. The ongoing exploration of this integration enhances the capabilities of laser systems and opens new avenues for innovation across various industries. The future of laser technology lies in its ability to adapt and respond intelligently to the demands of an ever-changing technological landscape, with automation and control systems playing a pivotal role in that transformation.

Advanced Laser Technologies

Advanced laser technologies transform industries through innovations that enable unprecedented precision, control, and power. These technologies include ultrafast lasers capable of femtosecond pulse durations, which are invaluable in applications ranging from delicate medical procedures to high-precision material processing. With the integration of machine learning, these technologies are becoming even more adaptive and efficient, pushing the boundaries of what lasers can achieve in modern applications.

Emerging Laser Technologies

Emerging laser technologies are transforming a wide range of fields, driven by advancements in machine learning and sophisticated laser science. These advancements have led to significant improvements in laser systems and are expanding their scope of application across scientific research, industrial manufacturing, medical treatments, and defense sectors. Machine learning empowers laser systems with refined control over essential parameters, including output power, wavelength tuning, beam shaping, intensity regulation, pulse duration, and energy output. As a result, lasers are becoming more efficient, adaptable, and precise, creating new possibilities for applications that previously seemed challenging or unattainable.

The advancement of continuous wave (CW) lasers, pulsed lasers, Chirped Pulse Amplification (CPA) systems, and fiber lasers highlight the technological evolution in laser science. CW lasers deliver steady, uninterrupted power that is ideal for applications requiring continuous beams, such as materials processing and medical procedures in some cases. Pulsed lasers, capable of delivering energy in controlled bursts, have opened doors for high-precision applications that require focused and intense energy delivery, such as micro-machining and laser eye surgery. CPA technology, which can amplify ultrashort pulses to high peak intensities without damaging optical components, has become foundational in high-energy physics and attosecond science. Fiber lasers, known for their efficiency and beam quality, are being adapted to various applications, including industrial machining and medical treatments.

Machine learning has introduced a transformative layer to these laser systems, enabling real-time optimization and adaptability. Machine learning algorithms can dynamically adjust laser parameters to match specific application requirements by analyzing vast datasets and utilizing predictive modeling. For example, machine learning in high-power fiber lasers allows for precision control of beam shaping and stability, improving the quality of applications such as cutting and welding. In CPA systems, machine learning can manage amplifier stages to reduce pulse distortion, optimize pulse duration, and maximize output power, resulting in more stable and higher-intensity pulses. This integration of machine learning within CPA systems facilitates advancements in high-intensity applications, such as high harmonic generation (HHG) and plasma generation, where precise pulse management is critical.

The synergy between machine learning and laser technology is particularly impactful in industrial manufacturing. Machine learning enables laser systems to make on-the-fly adjustments based on real-time data analysis, which is especially useful for cutting, welding, and engraving processes, where consistency and precision are essential. For example, in high-precision manufacturing sectors like microelectronics, automotive, and aerospace, machine learning can detect slight variations in material properties or environmental conditions and adjust laser parameters accordingly. This adaptability improves output quality, reduces waste, and enhances productivity. Additionally, machine learning algorithms enable lasers to automate quality control by detecting and correcting irregularities, ensuring consistency across large production runs.

Medical applications have also seen a substantial impact from laser technology improvements coupled with machine learning. Lasers are now integral to diagnostic and therapeutic procedures, enabling non-invasive and highly targeted treatments. Machine learning enhances the precision and personalization of laser-based treatments by analyzing patient-specific data, such as imaging scans, and predicting the ideal laser settings. In laser surgery for cancer treatment, machine learning can optimize parameters to ensure accurate targeting of diseased tissue while preserving healthy tissue. Ultrafast lasers, which produce pulses on the femtosecond timescale, offer minimal tissue damage and faster patient recovery, making them suitable for eye surgeries, skin treatments, and oncology procedures. In diagnostics, machine learning-enhanced lasers improve the accuracy and resolution of imaging, aiding in the early detection and precise localization of abnormalities, which is invaluable in fields like oncology, where early detection can significantly improve treatment outcomes.

Environmental applications of laser technology, empowered by machine learning, are becoming crucial tools for monitoring and conserving natural resources. Remote sensing lasers equipped with machine learning capabilities can perform atmospheric measurements to track pollutants, greenhouse gases, and deforestation, providing essential data for climate studies and environmental policy-making. For example, LiDAR systems integrated with machine learning can analyze large datasets to monitor land-use changes, assess water quality, and accurately detect pollution sources. The ability to interpret complex environmental data in real-time allows these laser systems to deliver actionable insights, supporting sustainable practices and promoting environmental conservation efforts. In scientific research, emerging laser technologies are invaluable in attosecond science, plasma physics, and particle acceleration. High harmonic generation (HHG), which involves converting fundamental laser frequencies to higher harmonics, enables researchers to observe ultrafast processes at attosecond timescales. Machine learning optimizes HHG setups by managing phase matching, improving efficiency, and ensuring harmonic stability, which is crucial for precise and repeatable experiments. Plasma generation, another key area of laser research, is pivotal in exploring nuclear fusion, astrophysical phenomena, and advanced particle acceleration techniques. With machine learning-based adjustments, high-power lasers enable stable plasma production by fine-tuning beam intensity and focus. The precision offered by machine learning enhances the reproducibility of these experiments, facilitating breakthroughs in fundamental physics and contributing to advancements in energy science.

Defense applications benefit from the precision and adaptability of machine learning-enhanced laser technologies. High-power lasers are increasingly used in directed-energy weapons, which rely on focused laser beams to disable aerial threats, including drones, missiles, and other projectiles. Machine learning enables these systems to track and adapt to moving targets, adjusting beam intensity, focus, and duration for maximum impact. This adaptability allows defense lasers to operate effectively in varied atmospheric conditions, enhancing their reliability in field operations. Beyond weaponry, machine learning-powered lasers are also employed in surveillance, where they assist in detecting and identifying objects over long distances. Infrared lasers enhanced with machine learning algorithms can distinguish between materials and track heat signatures, providing vital intelligence for security operations and threat assessment. The strategic integration of laser technology and machine learning strengthens national defense capabilities, offering real-time insights and accurate targeting solutions.

The role of machine learning within these systems is expanding across various domains. Machine learning's ability to process large volumes of data and make real-time adjustments transforms traditional laser systems into adaptable, intelligent tools. In industrial manufacturing, this adaptability fosters more efficient workflows, reduced waste, and higher productivity. In medical applications, machine learning-driven lasers enable more accurate, minimally invasive treatments, which improves patient outcomes and broadens the scope of laser-based therapies. Environmental monitoring benefits from the precision of machine learning, allowing for detailed ecological assessments that support sustainability efforts. In defense, machine learning enhances the efficacy of laser systems in threat neutralization and intelligence gathering.

The flowchart diagram begins with the evolution of laser technology, leading to different types of lasers on the left side. These laser types include CW lasers, used for applications that need a steady stream of light; pulsed lasers, useful for tasks requiring short bursts of energy; CPA systems, typically used in high-intensity applications; and fiber lasers, which are known for efficiency and versatility across various applications. The right side of the flowchart addresses the integration of machine learning in laser technology. This integration begins with model training and evaluation, where machine learning models are developed and assessed for their performance in controlling and optimizing laser parameters. After evaluation, models move into deployment, where they are used in real-world laser systems. Through continuous improvement, the models can make real-time adjustments to adapt to changing conditions or specific application needs. To maintain accuracy and adaptability, there is a feedback loop within the machine learning cycle, including data collection, analysis, and parameter optimization. This loop allows models to learn from new data, enhancing their performance.

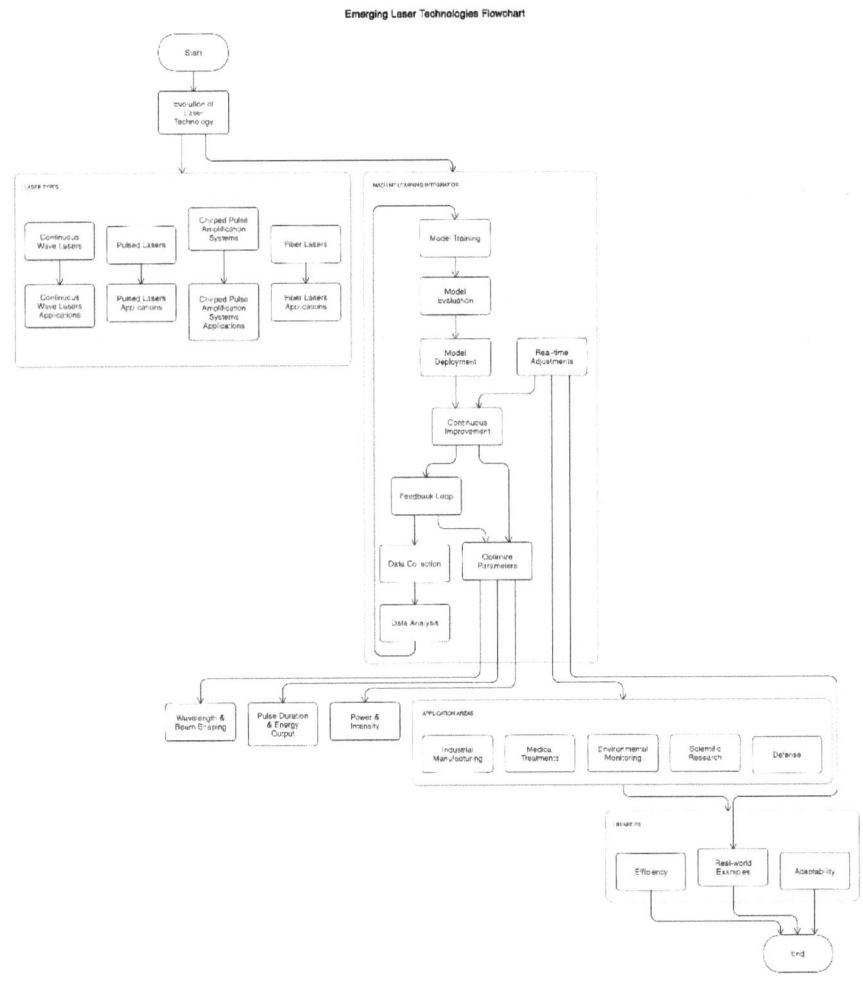

The lower section of the flowchart shows key laser parameters that can be controlled or optimized, including wavelength and beam shaping, pulse duration and energy output, and power and intensity. These parameters are crucial for tailoring laser performance to specific tasks. Finally, the flowchart lists various application areas where advanced lasers and machine learning are making an impact, such as industrial manufacturing for cutting, welding, and material processing; medical treatments in surgeries and therapies requiring precision; environmental monitoring for analyzing pollutants and atmospheric data; scientific research in experiments needing controlled light sources; and defense for applications requiring secure, high-power lasers. The final section highlights the benefits of integrating these technologies, such as increased efficiency, providing real-world examples of success, and adaptability in various fields, marking the end of the flowchart. This structured approach demonstrates how combining laser technology with machine learning drives efficiency, adaptability, and precision in diverse applications.

Integration of Artificial Intelligence (AI) with Advanced Lasers

Integrating AI with advanced laser technologies is reshaping the landscape of various scientific and industrial applications. Let's consider the flowchart diagram, which illustrates the integration of AI with advanced laser technologies. It starts with two main components: AI technologies (including machine learning algorithms and data analysis tools) and laser technologies (encompassing laser systems and parameters). These two components converge in various applications, such as industrial manufacturing, medical procedures, and materials processing. The integration leads to three key outcomes: optimized performance, enhanced precision, and predictive maintenance. These outcomes further result in benefits such as cost savings, improved safety, and the exploration of future directions, ultimately enabling the development of autonomous laser systems. All these factors combine to achieve an optimized and safer operation for advanced laser technologies. Overall, this flowchart highlights the journey from the start of AI and laser technologies to specific applications and outcomes.

Integration of AI with Advanced Laser Technologies

Ti-Sapphire-Based Chirped Pulse Amplifier System Integration with Machine Learning

Integrating a Ti-sapphire Chirped Pulse Amplifier (CPA) system with machine learning can significantly enhance performance by automating optimization tasks, improving pulse shaping, and adapting to dynamic conditions. CPA is a technique developed to amplify ultrashort laser pulses to high power levels without damaging the optical components. The CPA system temporarily stretches a short laser pulse using a dispersive element (often a grating stretcher), reducing its peak power. The stretched pulse is then amplified through one or more laser amplifiers and recompressed to its original femtosecond duration. CPA systems typically include several essential components: a pulse stretcher, multiple amplifiers (often including regenerative and multipass amplifiers), and a pulse compressor. A stable oscillator also serves as the initial seed pulse source. The system must control several parameters precisely to achieve optimal performance, including pulse duration, energy, power, center wavelength, and spectral width; the block diagram below shows the CPA process.

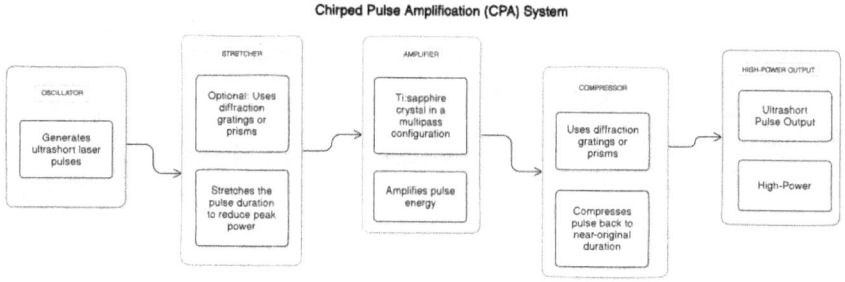

In a Ti-sapphire CPA, precise pulse compression is crucial to achieving ultrashort, high-intensity pulses. Traditional methods require manual tuning of compressor gratings, which can be sensitive to environmental factors. Machine learning algorithms can analyze real-time pulse data to optimize pulse compression settings, minimizing pulse duration by dynamically predicting the ideal grating positions or fine-tuning dispersion compensation. Fluctuations in the Ti-sapphire gain medium due to thermal effects or varying pump power can lead to inconsistent output pulse energies and beam quality. A machine learning model trained on system data (e.g., pump power, pulse energy, and beam profile) can actively control the pump laser and other amplifier parameters to maintain stable output. Predictive models could also forecast the onset of gain saturation or thermal effects and adjust system parameters preemptively. Maintaining the phase of the amplified pulse, especially in multi-pass systems, is essential to preserve temporal coherence, but it can be disturbed by environmental noise or drifts.

Machine learning can be applied to feedback loops that monitor and stabilize the phase of the amplified pulse. Neural networks or reinforcement learning could analyze the phase changes and adjust the system's phase compensation units in real-time to ensure stable output. Nonlinear effects such as self-phase modulation (SPM) or Kerr lensing can distort the pulse shape during amplification, especially at high intensities. By analyzing the system's nonlinear behavior, a machine learning model could predict the onset of unwanted nonlinear effects and make real-time adjustments to minimize their impact (e.g., by controlling pulse energy or adjusting optical elements). Over time, maintaining the alignment and performance of a Ti-sapphire CPA requires constant monitoring and frequent manual intervention. Machine learning algorithms can track long-term system performance, predict component degradation, and schedule maintenance based on predictive diagnostics, reducing downtime and ensuring optimal system performance. Tuning a CPA system for different pulse characteristics (e.g., energy, duration, spectral bandwidth) is manual. Using a self-learning algorithm like reinforcement learning, the system could automatically explore different parameter configurations and optimize for the desired pulse characteristics, e.g., shortest pulse duration or highest possible energy based on user input.

In precision laser micromachining, a Ti-sapphire laser with machine learning integration can optimize pulse energy, beam profile, and focal spot size to achieve ultra-high precision cuts or ablations. In medical applications like eye surgery or cancer treatment, machine learning can help fine-tune pulse duration and energy to ensure safe and efficient tissue ablation. A machine-learning CPA system can optimize pulse parameters for effective laser-induced breakdown spectroscopy (LIBS) or other diagnostic techniques for environmental applications like remote sensing or atmospheric studies. Integrating machine learning into Ti-sapphire CPA systems can make these lasers smarter and more efficient.

Empowering Laser Technology with Machine Learning

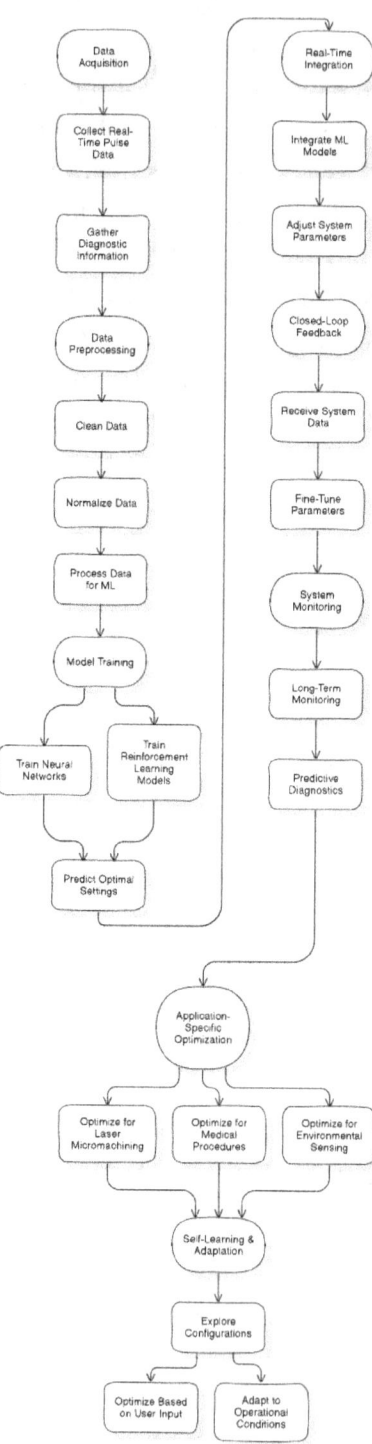

The flowchart diagram outlines a process for optimizing the Ti-sapphire CPA laser system using machine learning techniques. It begins with data acquisition, where real-time pulse data and diagnostic information are collected and preprocessed through cleaning and normalization. The processed data trains neural networks and reinforcement learning models to predict optimal laser system settings. These models are integrated into the real-time system, adjusting system parameters based on feedback and fine-tuning them through closed-loop monitoring.

The system undergoes long-term monitoring with predictive diagnostics. Application-specific optimizations are made for tasks like micromachining, medical procedures, or environmental sensing, enabling the system to self-learn and adapt to changing conditions or user inputs for enhanced performance.

How to integrate machine learning into a Ti-sapphire CPA system:
The CPA system involves a series of essential steps that range from data acquisition to real-time system control. Initially, relevant real-time data from the CPA system, such as pulse characteristics, amplifier parameters, environmental conditions, and performance metrics, are gathered using sensors and diagnostic tools. This data is then preprocessed to clean, normalize, and reduce dimensionality to make it suitable for machine learning models. Machine learning models are selected and trained for specific tasks like pulse compression optimization, adaptive gain control, phase stabilization, nonlinear effects prediction, and system diagnostics. For example, supervised learning algorithms can optimize pulse compression, while reinforcement learning can dynamically adjust gain control based on real-time system data. Once trained, these models are integrated into the CPA system for real-time control. This involves creating control interfaces that allow machine learning models to interact directly with the system and continuously implementing closed-loop feedback to adjust system parameters in response to changing conditions. After integration, the system is rigorously tested to ensure the models optimize performance effectively and remain stable under varying conditions. Over time, these machine learning models can be maintained and updated through online learning, allowing the system to adapt as new data is gathered and system conditions evolve. For instance, a neural

network could be trained to predict the optimal compressor grating position for pulse compression, which is then continuously refined through real-time feedback, ensuring minimal pulse duration even in fluctuating environments. Ultimately, integrating machine learning into the CPA system automates complex optimization tasks, improves system stability, and enhances overall efficiency, allowing the system to adapt to operational changes dynamically.

Integration of ML into Ti-Sapphire CPA System

The block diagram illustrates that integrating machine learning into a Ti-sapphire-based CPA system begins with real-time data acquisition, including pulse characteristics and environmental conditions. This data is preprocessed, cleaned, normalized, and stored in a historical database. Machine learning models are then selected, trained using this historical data, and integrated into the system for real-time control, where closed-loop feedback enables continuous adjustment of system parameters based on the machine learning model's predictions. After initial testing, the system undergoes tuning to optimize performance, ensuring the models effectively control pulse compression, gain, and spectral and temporal phase stabilization. Long-term model maintenance allows for ongoing adaptation, where the models are periodically updated with new data to maintain optimal system performance as conditions evolve.

CASE STUDIES OF CONVERGENCE OF MACHINE LEARNING AND LASER TECHNOLOGY

Integrating machine learning techniques into laser research and industrial applications reshapes how lasers are designed, controlled, and utilized. This chapter presents several case studies that illustrate the transformative impact of machine learning on laser systems, highlighting their practical applications and the advancements they enable.

Case Studies of Machine Learning Enhancing Laser Technology

Optimization of Laser Cutting

One significant case study uses machine learning algorithms to optimize laser cutting processes. Implement predictive models that analyze material type, thickness, and laser power settings. By employing a supervised learning approach, predicting optimal cutting speeds and feed rates reduces material wastage and improves production efficiency.

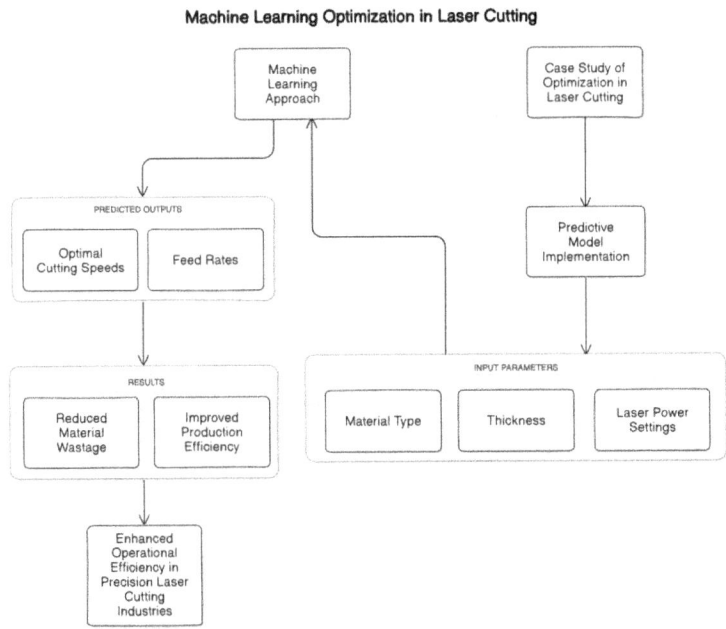

The diagram showcases the use of machine learning to optimize laser cutting processes. It begins with a machine learning approach and a case study in optimization, leading to implementing a predictive laser-cutting model. The predictive model uses input parameters such as material type, thickness, and laser power settings to determine the optimal cutting speeds and feed rates. These predicted outputs result in reduced material wastage and improved production efficiency. This case highlights how machine learning can enhance operational efficiency in industries reliant on precision laser cutting.

Decision Support System for Skin Lesion Analysis

Another interesting and convincing example is laser-based medical treatments, particularly dermatology.

Development of a Decision Support System for Skin Lesion Analysis

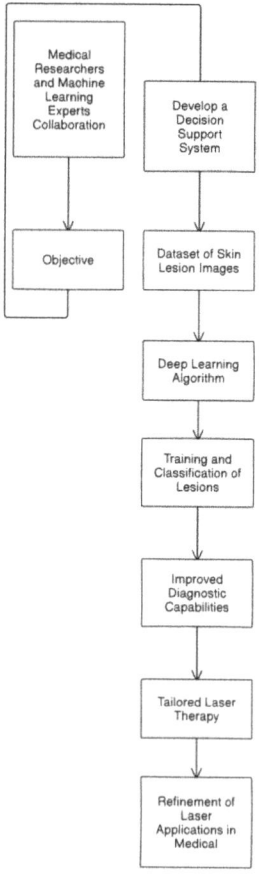

A decision support system developing that utilizes deep learning algorithms to analyze skin lesions. The system was trained on a vast dataset of images, enabling it to classify lesions with remarkable accuracy. Implementing this technology has improved diagnostic capabilities and personalized treatment plans, allowing medical professionals to tailor laser therapy to individual patient needs. The block diagram shows how it may go; it begins with collaboration between medical researchers and machine learning experts, creating a deep learning-based system trained on a large dataset. The system enhances diagnostic accuracy, enables tailored laser therapy, and ultimately refines laser applications in medical treatments.

Optimization of Laser Welding

In laser materials processing, reinforcement learning is important in optimizing laser welding parameters to meet the demands of modern industries. Machine learning models can be trained to adaptively adjust critical variables, such as laser intensity and focal position, in real time by leveraging simulations of various welding scenarios. This approach ensures higher weld quality, improved consistency, and reduced defects. The process begins with an exploration phase, where engineers experiment with reinforcement learning methods to understand their application to laser welding. Objectives are to enhance weld quality, minimize material waste, and improve efficiency. Key welding parameters like laser intensity and focal position are optimized through computational simulations of diverse welding scenarios. These simulations provide data to train a reinforcement learning model capable of real-time decision-making.

The trained model dynamically adjusts laser parameters during welding, responding instantly to changing conditions. This results in consistent welds of higher quality, reducing defects and increasing efficiency. Over time, the system evolves into a smarter, more responsive welding process that adapts to complex industrial requirements in aerospace, automotive, and advanced manufacturing sectors.

This dynamic approach, as demonstrated in the diagram, highlights how machine learning can lead to innovative solutions in laser technology. By integrating reinforcement learning into laser welding, the process becomes more adaptive, precise, and efficient, meeting the highest standards of modern industrial applications.

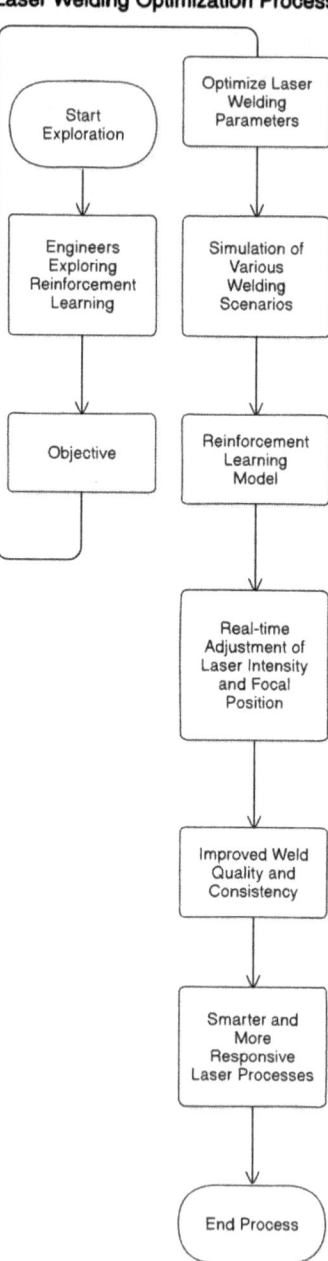

Laser Imaging

Machine learning advancements have significantly transformed laser-based imaging techniques, especially in biological research. The application of **Convolutional Neural Networks (CNNs)** has enabled the enhancement of resolution and accuracy in laser-induced fluorescence imaging. By training these networks on extensive fluorescence image datasets, researchers have refined imaging systems to detect and analyze fine details in cellular structures and biochemical processes that were previously difficult to observe.

The block diagram illustrates this process step-by-step, starting with improving the resolution and accuracy of laser imaging techniques. This objective sets the foundation for the implementation of advanced machine learning models. The next step involves using CNNs, a deep learning algorithm specifically designed for image recognition and analysis. These networks are particularly well-suited for identifying patterns in complex biological images. Following this, the system undergoes training with extensive fluorescence image datasets. These datasets consist of high-quality images of biological samples captured using laser-induced fluorescence techniques. The training process allows the CNN to learn and extract meaningful features from the images, such as cellular structures, molecular arrangements, and dynamic biological changes.

As a result of this training, the imaging system achieves enhanced resolution and accuracy. This means the imaging process becomes more reliable and precise, allowing for the visualization of structures and processes previously indistinct or beyond the resolution limits of traditional imaging methods.

Finally, these improvements lead to deeper insights into biological processes, enabling researchers to study cellular mechanisms, molecular interactions, and other intricate biological phenomena in greater detail. This progression improves the quality of research and opens new avenues for discoveries in fields such as molecular biology, medicine, and drug development.

The block diagram encapsulates how machine learning, particularly CNNs, is integrated into the laser imaging process. It demonstrates a clear flow, from identifying the objective to achieving transformative outcomes in biological research.

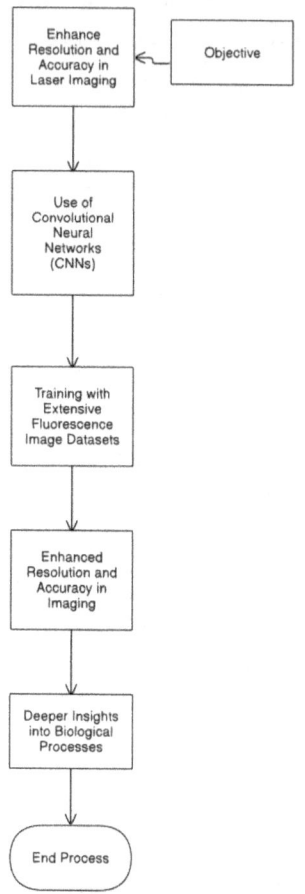

Advancement in Machine Learning for Laser Imaging

Laser Communication

Laser communication has seen remarkable innovations through machine learning. Machine learning algorithms to optimize signal processing in free-space optical communication systems. The algorithms were designed to mitigate the effects of environmental disturbances, such as turbulence and scattering. The system demonstrated increased data throughput and reliability by continuously adapting to changing conditions. This case reinforces the idea that machine learning enhances the performance of laser technologies and paves the way for new applications in fields like telecommunications, where effective data transmission is critical. The diagram below illustrates the process of using machine learning to enhance laser communication systems. It starts with research and development by applying machine learning algorithms to address environmental disturbances. The system adapts to changing conditions, improving data transmission performance.

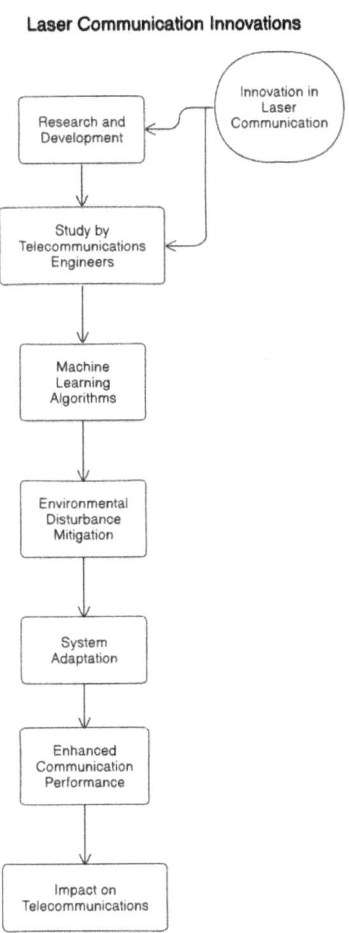

Laser Communication Innovations

Special Case Studies of Machine Learning Enhancing Laser Technology

Optimizing Chirped Pulse Amplification System

Chirped Pulse Amplification (CPA) systems typically include several essential components: a pulse stretcher, multiple amplifiers, and a pulse compressor. A stable oscillator also serves as the initial seed pulse source.

The system must precisely control several parameters to achieve optimal performance, including pulse duration, energy, power, center wavelength, spectral width, etc. Machine learning has emerged as a promising tool for optimizing various parameters within CPA systems, providing data-driven solutions to complex, interdependent issues. By training on operational data and learning relationships among variables, machine learning models can optimize parameters like energy, beam profile, and pulse duration, all in real-time. This approach eliminates labor-intensive trial-and-error optimization and enables the system to adapt dynamically to environmental conditions or input laser characteristics fluctuations.

Firstly, the stability of the oscillator, which provides the initial seed pulse for the CPA system, is fundamental to the entire process. Machine learning can monitor the oscillator's output in terms of pulse shape, center wavelength, and peak power stability, making small adjustments to maintain a consistent seed pulse. This stability cascades through the amplification chain, resulting in a more stable and reliable overall output. If they are in the CPA system chain, machine learning can be implemented in CPA systems to optimize amplifier parameters. These amplifiers play a crucial role in scaling the energy and peak power of the laser pulse. Machine learning algorithms, such as reinforcement learning or neural networks, can analyze data on previous amplification stages to predict optimal pump power levels and beam profiles that maximize energy gain while minimizing distortions. By continuously monitoring fluctuations in the laser parameters and adjusting the pump laser power or cooling systems, the machine learning model helps maintain a stable output, even under varying operational conditions.

Machine learning can also assist in fine-tuning the output spectral width and central wavelength of the laser pulse, which are crucial for achieving high peak power upon recompression. These parameters often drift due to environmental factors or changes in the amplifier gain spectrum, leading to suboptimal pulse compression. Machine learning algorithms can analyze the output spectra in real-time and make minor adjustments to amplifier settings or feedback to the oscillator, ensuring that the center wavelength remains stable and the spectral width is optimized.

Furthermore, the algorithms can modify the beam profile by adjusting beam-shaping optics, contributing to more uniform intensity distribution across the beam and reducing hot spots that could damage optical components.

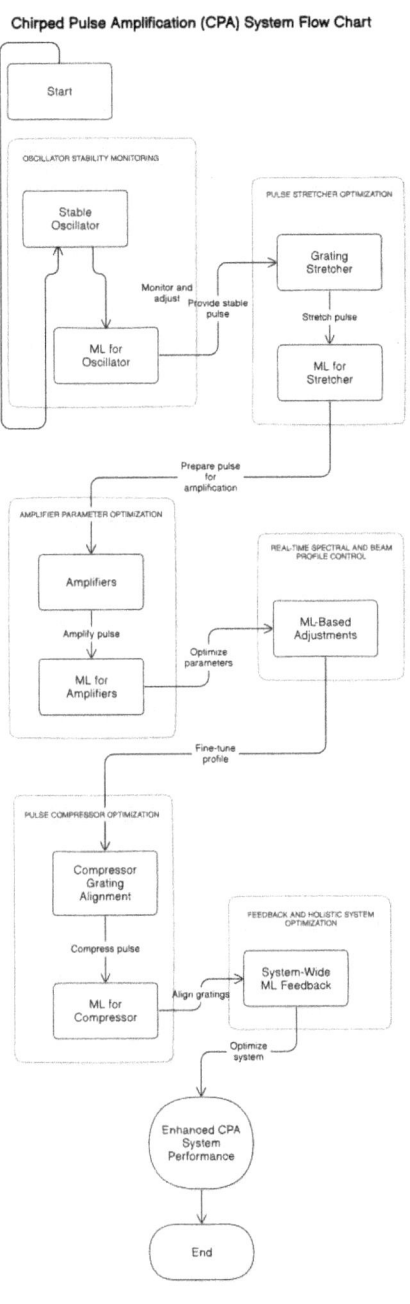

Another critical application of machine learning in CPA systems is optimizing the pulse stretcher or compressor gratings. The precise alignment and spacing of gratings are vital for achieving the shortest pulse duration after amplification. Machine learning algorithms can model the relationship between compressor grating separation and pulse duration, automatically adjusting grating positions to account for minor misalignments or thermal expansion that could otherwise degrade pulse quality. This real-time adjustment enhances pulse compression and maintains the integrity of the beam profile. Machine learning algorithms can also track feedback from various stages in the CPA system, allowing for holistic optimization across all components.

The self-explanatory diagram shows the role of machine learning in optimizing a CPA system and how machine learning monitors and adjusts key stages of the system. Each stage uses machine learning to enhance specific parameters, such as stability, pulse duration, and beam quality, ensuring a stable, high-power, ultrashort pulse output.

Integrating machine learning into CPA systems enables unprecedented control over complex, dynamic parameters, from pulse energy and power to spectral characteristics and pulse duration. Machine learning enhances the stability and performance of CPA systems by optimizing components such as the pump laser, amplifier stages, beam profile, and grating separation. This technology-driven approach improves the efficiency and adaptability of CPA systems. It unlocks new possibilities for high-power, ultrashort pulse generation.

Machine learning algorithms in optimizing CPA systems: Integrating machine learning algorithms into optimizing CPA systems requires selecting and configuring algorithms that can address the system's multi-dimensional, complex parameter space. Given the dynamic interactions between variables such as pulse energy, power, duration, and stability, the choice of a machine learning model depends on the specific optimization task within the CPA system. Machine learning has proven particularly effective in optimizing the pump power, amplifier gain, spectral characteristics, and overall system stability, enabling real-time adjustments and performance improvements that are difficult to achieve manually. **Reinforcement Learning (RL)** is a suitable approach for managing real-time control and optimization in CPA systems, especially for tasks requiring continuous adaptation, such as adjusting pump power, optimizing amplifier gain, and stabilizing pulse energy. In reinforcement learning, an "agent" learns to make decisions by receiving feedback as rewards or penalties from the environment (i.e., the CPA system) based on its actions. A reinforcement learning model, such as a Deep Q-Network (DQN) or Proximal Policy Optimization (PPO), can be trained to maximize a reward function that considers pulse energy stability, minimized distortions, and peak power. The reinforcement learning agent adjusts parameters like pump laser power, beam profile, or cooling levels in the amplifiers, seeking to maximize energy gain and maintain pulse consistency. This adaptability is ideal for managing fluctuations in operational conditions and maintaining stability across the amplification chain. **Artificial Neural Networks (ANN)**, specifically deep learning architectures, are also effective for optimizing spectral characteristics like the output pulse's central wavelength and spectral width. In CPA systems, these parameters directly impact pulse compression and peak power after amplification. Variations in amplifier gain and environmental factors can cause drift in the wavelength or narrowing of the spectral width, resulting in suboptimal pulse duration upon recompression. A **Convolutional Neural Network (CNN)** or a **Recurrent Neural Network (RNN)** could be trained on a large dataset of spectral and temporal measurements from the CPA system under various operational conditions. By predicting the optimal values of parameters that maintain spectral characteristics, the network can assist in real-time adjustments to amplifier settings or oscillator parameters, helping the system automatically counteract drift and enhance pulse compression. **Genetic Algorithms (GA)** offer a robust solution for multi-objective optimization problems, where the goal is to balance conflicting objectives, such as maximizing pulse energy while minimizing phase distortions or managing nonlinear effects. In a CPA system, a genetic algorithm can be used to determine the optimal configuration of amplifier parameters, including the number of amplifier stages, pump power levels, and grating separation in the stretcher/compressor. Each configuration is evaluated based on a fitness function that considers multiple objectives, such as energy stability and pulse shape, and minimizes nonlinear distortions. The genetic algorithm is then run on a high-performance computing system to simulate multiple configurations and converge on an optimized parameter set for the CPA system, making it particularly useful during initial system setup or when reconfiguring for different applications. **Support Vector Machines (SVM)** can assist in classification tasks within CPA systems, particularly in identifying patterns or anomalies in beam profiles and ensuring system stability. SVM can classify the system's operational state based on real-time data, determining whether the beam profile is within acceptable limits or if corrective action is needed. SVM can be trained on a dataset of beam profile images labeled with categories such as "stable," "unstable," or "misaligned." Once trained, the SVM model can classify the incoming beam profiles in real-time, triggering adjustments if instability or misalignment is detected, thereby maintaining uniformity in beam intensity distribution and protecting optical components. **Bayesian Optimization** is an efficient approach for optimizing parameters that require a trade-off between performance and precision, making it ideal for tasks like adjusting grating separation to achieve optimal pulse duration. Optimizing grating separation is critical for minimizing pulse duration post-amplification. Small misalignments or temperature-induced shifts in grating positions can lengthen the pulse and reduce peak power. Bayesian optimization can be used to find the optimal grating separation distance by evaluating various configurations and constructing a model of how separation distance affects pulse duration, minimizing the number of adjustments needed to reach optimal alignment.

Day-to-Day Laser Parameter Optimization and Stability

Focusing on integrating machine learning in day-to-day laser parameter optimization and stability, a laser facility can implement advanced algorithms to enhance its operational processes. Traditionally, adjusting parameters such as power, pulse duration, beam profiling, beam pointing stability, and focus settings require extensive manual experimentation, leading to inefficiencies and inconsistencies. The laser facility can analyze historical performance data from various laser operations by leveraging machine learning techniques. For instance, supervised learning algorithms can be trained on datasets encompassing previous outcomes under varying conditions, enabling the prediction of optimal parameter configurations for specific tasks. Additionally, reinforcement learning can facilitate real-time adjustments based on feedback from ongoing operations, ensuring stability and consistency in laser performance.

Moreover, machine learning models can actively monitor laser stability by analyzing fluctuations in output and identifying potential issues before they escalate. This proactive approach can minimize downtime and enhance the precision of laser applications. Unsupervised learning algorithms can be employed to detect patterns and anomalies in laser performance, such as fluctuations in output power or changes in beam quality. By dynamically adjusting parameters, the system can maintain optimal performance under varying conditions, potentially substantially reducing equipment failures and increasing operational efficiency. Overall, this case study underscores the transformative potential of machine learning on laser technology, paving the way for smarter, more reliable, and more efficient laser systems in industrial and fundamental and applied laser research applications.

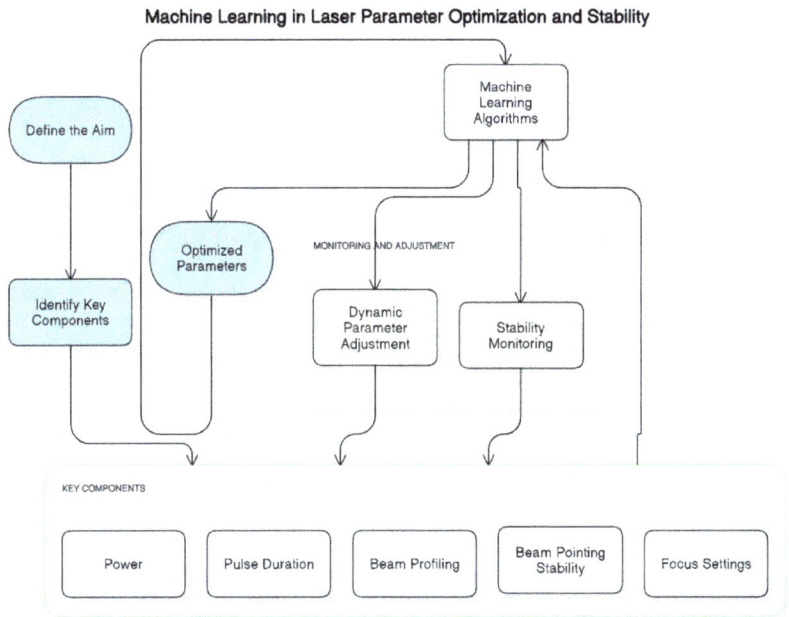

The conceptual block diagram shows the laser optimization and stability case study. The process starts with defining the aim and identifying key components involved in laser operation, including parameters like power, pulse duration, beam profiling, beam pointing stability, and focus settings. These components are fed into machine learning algorithms, which analyze and optimize these parameters to achieve stable and efficient laser output. A continuous monitoring and adjustment cycle, consisting of dynamic parameter adjustment and stability monitoring blocks, allows real-time tuning based on ongoing feedback. The output is a set of optimized parameters that ensure consistent laser performance, improving efficiency and reliability in day-to-day operations.

DATA-DRIVEN APPROACHES IN LASER TECHNOLOGY

Data collection and analysis are crucial in advancing laser technology, particularly when integrated with machine learning techniques.

Importance of Data Collection and Analysis

In research and development, collecting robust data allows identifying patterns, validating hypotheses, and driving innovation. Starting from day-to-day data acquisition of laser systems operations, the diverse applications of laser technology, ranging from telecommunications to medical procedures, necessitate a comprehensive understanding of how lasers behave and interact with various materials and environments. We can make informed decisions that lead to more effective laser designs and applications by systematically gathering data on operations and these interaction applications.

Machine learning enhances the capability of data analysis, allowing for the extraction of meaningful insights from vast datasets. With traditional analytical methods, the complexity and volume of data generated in laser experiments can be overwhelming. However, machine learning algorithms can process this data more efficiently, uncovering relationships and trends that may not be immediately apparent. This capability is particularly important in optimizing laser parameters such as intensity, wavelength, beam profiling, pulse duration, and focusing, which can significantly impact the outcomes of laser applications in fields like manufacturing and medical.

The importance of data collection extends beyond the laboratory; it also encompasses real-world applications and user feedback. Gathering data from end-users can provide invaluable insights into the performance and reliability of laser systems. This feedback loop enables engineers to refine their designs based on actual usage patterns and challenges faced in the field. Moreover, integrating user-generated data into machine learning models can enhance predictive maintenance, ensuring laser systems operate at peak efficiency and reduce downtime. As the landscape of laser technology evolves, data collection and analysis become increasingly prominent. Emerging technologies, such as adaptive optics and real-time monitoring systems, generate a continuous data stream that can be harnessed for ongoing improvements. This data can enhance current technologies and anticipate future trends and challenges. Laser technology can remain at the forefront of innovation by fostering a data-driven decision-making culture. The synergy between data collection, analysis, and machine learning is vital. The insights gained from well-structured data collection can lead to breakthroughs in laser technology and applications.

Big Data and Its Role in Laser Research

Big data has emerged as a primary component in the evolution of laser research, offering unprecedented opportunities for innovation and efficiency. The vast amounts of data generated from experiments, simulations, and real-time measurements have become integral to understanding complex laser phenomena. We can analyze this data to identify patterns, optimize laser designs, and improve operational efficiencies. With advanced data analytics, meaningful conclusions can be drawn from millions of data points, leading to faster advancements in laser technology.

One of the primary applications of big data in laser research is in the area of predictive modeling. By harnessing large datasets, we can create models that predict the behavior of lasers under various conditions. These models can include temperature fluctuations, material properties, and environmental influences. The ability to predict outcomes accurately helps design more efficient lasers and minimizes trial-and-error approaches in experimental setups. Consequently, this accelerates innovation and reduces costs associated with research and development. Moreover, machine learning algorithms play a significant role in analyzing big data in laser technology. These algorithms can process vast datasets far more efficiently than traditional statistical methods. For instance, machine learning can optimize laser parameters for day-to-day operation for specific applications, such as laser application research, HHG, plasma wakefield, etc., and industrial applications, such as cutting and welding or medical treatments. By training models on historical data, we can significantly enhance the precision and effectiveness of laser operations.

Integrating big data analytics with laser research also facilitates enhanced diagnostics and monitoring. Real-time data acquisition from laser systems allows continuous monitoring of performance and detect anomalies. This capability is crucial for applications that require high precision. By identifying issues as they arise, implement corrective measures promptly, ensuring that laser systems operate at peak efficiency and reliability.

Case Studies on Data-Driven Laser Innovations

Data-driven innovations in laser technology have significantly transformed various applications, showcasing the synergy between advanced algorithms and laser systems. This subchapter delves into several case studies illustrating how data-driven machine learning techniques are applied to enhance laser performance, optimize processes, and enable new capabilities. From industrial manufacturing to medical applications, the integration of machine learning has streamlined operations and opened avenues for unprecedented advancements in laser technology.

One prominent case study involves using **machine learning algorithms to optimize laser cutting processes from historical data** in manufacturing. The predictive model is implemented to analyze parameters such as laser intensity, power, focusing and cutting speed, and material type. By feeding the model with historical data, fine-tune the laser settings in real-time, improving cutting precision and reducing material waste. The outcome enhanced productivity and lowered operational costs, demonstrating the significant impact of data-driven methodologies in industrial applications.

The block diagram shows it starts by gathering historical data and feeding it into a predictive model. The model considers various parameters, including laser intensity, power, cutting speed, material type, and real-time fine-tuning. These inputs help the model optimize the cutting process, improving cutting precision and reducing material waste, demonstrating how machine learning can streamline and optimize laser-cutting operations.

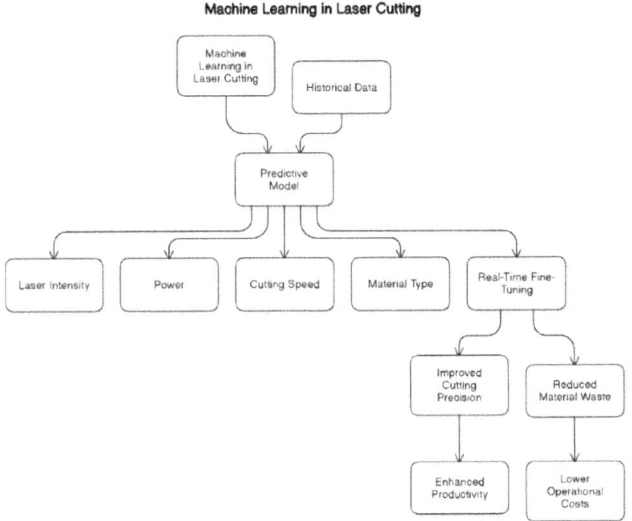

In the **medical field**, another case study highlights the **application of machine learning in laser-assisted surgeries**. It has developed an intelligent system that analyzes patient-specific data, including tissue type and lesion morphology, to determine optimal laser parameters for surgical procedures.

The diagram outlines a machine learning model, which begins with patient data collection and is followed by data preprocessing to prepare the data for analysis. The preprocessed data is used to create a data-driven model, which undergoes machine learning model analysis to identify patterns and insights relevant to laser surgery. Based on the analysis, the system optimizes laser parameterizations to fine-tune the laser settings for individual patients. These optimized parameters are then applied to the laser system. During surgery, monitoring and feedback are crucial for real-time adjustments, forming a feedback loop that continuously improves model accuracy and laser control. Finally, patient recovery and data logging capture post-surgery outcomes, which can be used to refine the model further for future cases. This process ensures a data-driven, patient-specific approach to laser surgery. The success of this project underscores the potential of machine learning to revolutionize medical laser applications by tailoring interventions to individual patient needs.

Machine Learning Model for Laser Surgery

```
Patient Data Collection
        ↓
Data Preprocessing
        ↓
Data-Driven Model Development ←─┐
        ↓                        │
Machine Learning Model Analysis  │
        ↓                        │
Laser Parameter Optimization     │ Feedback Loop
        ↓                        │
Laser System                     │
        ↓                        │
Monitoring and Feedback          │
        ↓                        │
Patient Recovery and Data Logging ┘
```

Developing novel laser materials and structures has benefited from data-driven insights. One case study explored the **use of machine learning to predict the properties of new laser gain media**. By analyzing a vast database of material compositions and their corresponding laser performance metrics, trained models can identify promising candidates for high-efficiency lasers.

The block diagram starts from the initial stage. The first phase involves material collection and data processing, where materials are gathered, and the collected data is preprocessed for further analysis. This information feeds into a model development phase, where machine learning techniques create and analyze data-driven models. After the model is developed, it progresses to the experimental phase, where experiments are conducted based on the model's insights.

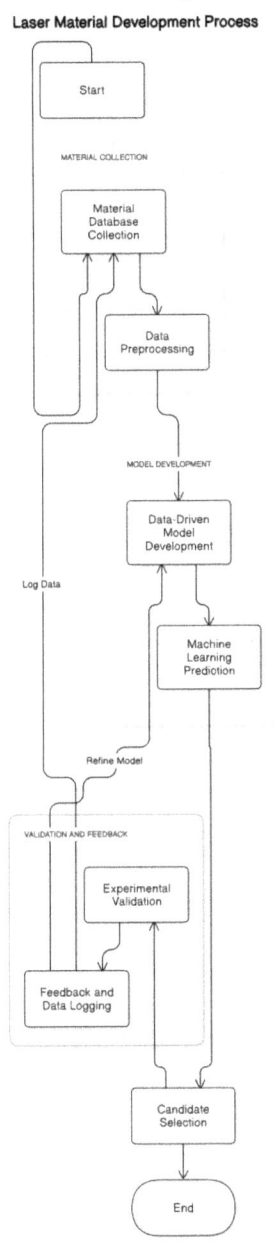

The feedback and data logging stage records the results and observations from these experiments. This feedback loops back to the model, enabling continuous refinement of the parameters. The process concludes with a decision stage where final considerations are made before the end of the workflow. This cyclical approach ensures iterative improvements in laser material development through continuous learning and feedback. Combining **machine learning with laser-based sensing technologies** presents another compelling case.

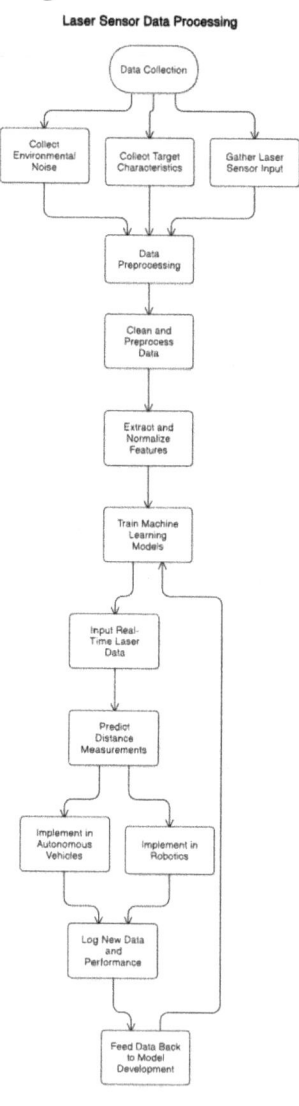

A study focused on deploying machine learning algorithms to enhance the accuracy of laser-based distance measurements in various environments. Training models on diverse datasets that included environmental noise and target characteristics improved the reliability of laser sensors in challenging conditions. This advancement has implications for autonomous vehicles and robotics applications, where precise distance measurement is critical. These case studies illustrate the transformative power of data-driven innovations in laser technology.

The diagram illustrates a laser sensor data processing workflow, starting with data collection from various sources, including environmental noise, target characteristics, and laser sensor input. After data is gathered, it goes through preprocessing steps such as cleaning, preparing, and feature extraction to make it suitable for analysis. The preprocessed data is then used to train a machine learning model. Once trained, the model receives real-time data inputs to predict and provide measurements. These predictions are implemented in applications across avionics and robotics. The model's performance is logged based on new data, which is fed back into the development process to refine further and improve the model. This cyclical process ensures continuous enhancement of the laser sensor system through ongoing learning and adaptation.

In conclusion, four diagrams of the four case studies illustrate data-driven machine learning models to improve laser-based technologies in distinct fields. In each case, the workflow begins with data collection, which forms the basis for developing and refining machine learning models. These systems continuously improve accuracy and performance by leveraging vast datasets and feedback loops. Each diagram underscores the role of feedback from real-world applications, allowing continuous refinement of the machine learning models to meet specific technological challenges.

Special Case Study on Data-Driven Machine Learning

Machine Learning Enhances Laser System Stability Through Data-Driven Optimization in Day-to-Day Operation

A special case study of a data-driven laser system optimization reveals how continuous monitoring and predictive algorithms can stabilize day-to-day operations. Integrating machine learning models with real-time data allows systems to adjust critical parameters such as laser power, alignment, beam profiling, and cooling systems to ensure consistent performance. Operators can feed historical data, including temperature fluctuations, beam output, and previous adjustments, into a model that predicts necessary tweaks to prevent degradation or misalignment, resulting in more stable and efficient laser operation.

For instance, a laser system used daily for precise tasks such as laser application research or engraving and cutting can exhibit variations in output due to ambient temperature changes or component wear. The model analyzes past system behavior under different conditions and adjusts real-time settings to mitigate performance drops. This proactive approach reduces the need for constant manual recalibration and improves operational consistency, saving time and extending the lifespan of the laser components.

In addition to optimizing performance, the system can detect signs of instability, such as small drifts in beam alignment or power output variations, that may go unnoticed by the operator. Early detection allows the model to recommend specific maintenance tasks or adjustments, preventing larger issues that could disrupt operations.

The flowchart diagram represents a sequence of processes optimizing a data-driven laser system. It begins with collecting real-time and historical data fed into a data collection system. A machine learning model then processes this data to identify trends, performance metrics, and areas for improvement. Based on the model's insights, critical parameters of the laser system are adjusted. These parameters include the cooling systems to manage heat, the alignment of the laser for precision, laser power control, and performance metrics such as efficiency and stability. Adjustments to these parameters impact the laser system's beam output, stability, and overall performance. The process includes a feedback loop, where the system continuously collects data, makes adjustments, and feeds new information into the model. Maintenance recommendations are also provided by the model based on system performance. As the system stabilizes and the beam output is optimized, the process moves towards completion, ensuring the laser operates at peak efficiency through continuous improvements. This case study shows that data-driven optimization enhances precision and stability, ensuring day-to-day operations run smoothly.

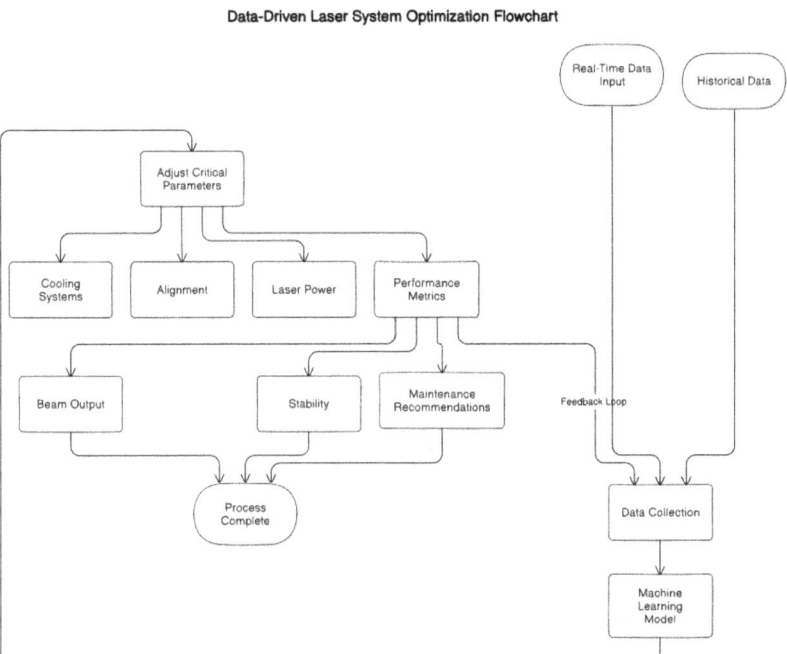

Data-Driven Laser System Optimization Flowchart

Optimization and Stability of Ti-Sapphire CPA Laser System with a Data-Driven Machine Learning

A data-driven approach in Ti-sapphire-based CPA laser systems leverages real-time data analysis, machine learning, and historical data to optimize and stabilize critical components of the CPA system, such as the oscillator, stretcher, amplifier, and compressor. These components are sensitive to temperature, alignment, and power fluctuations, affecting pulse stability, peak power, and overall system performance. A data-driven approach can spot small patterns and changes that suggest potential instability or suboptimal performance by constantly monitoring and analyzing data from these components. This allows for adjustments and fixes before problems worsen, keeping everything running smoothly and efficiently. For the **oscillator**, the seed source of the CPA system, stability in frequency and mode-locking (for detail please see the further reading and references) is important. Minor fluctuations in environmental conditions, like temperature or vibration, can cause frequency drift or mode-locking noise, ultimately impacting pulse duration and shape. A data-driven approach can monitor the oscillator's output in real-time, tracking parameters such as pulse timing jitter and frequency stability. Machine learning models trained on historical data can predict deviations based on current trends, allowing for real-time adjustments. For example, suppose the model detects an instability in the mode-locking or phase locking of the oscillator. In that case, it can signal adjustments to the cavity length or feedback controls to re-stabilize the oscillator, thus preserving the quality of the initial pulse entering the amplification stage. A critical component following the oscillator in the CPA system is the **stretcher**, which expands the ultrashort seed pulse to a longer duration to reduce its peak power, making it safe for amplification. The stretcher typically uses a diffraction grating to introduce controlled dispersion, elongating the pulse to the picosecond or nanosecond range. In a data-driven approach, real-time pulse duration and dispersion characteristics data are analyzed to maintain optimal stretching conditions. Even minor misalignments or environmental changes, such as thermal shifts, can alter the dispersion and impact pulse stretching. Machine learning models can detect suboptimal stretching patterns, prompting automatic adjustments to gratin alignment to maintain consistent pulse duration. This ensures that the pulse entering the amplifier is within the designed parameters, safeguarding the downstream components and enhancing the overall stability of the CPA laser system. The **amplifier** stage, particularly in multi-pass configurations, is where the pulse energy is significantly boosted. However, this stage is highly sensitive to thermal effects and alignment stability. Temperature changes in the Ti-sapphire crystal due to intense pumping or malfunctioning of the cooling system can lead to thermal lensing, causing beam distortion and loss of efficiency. The data-driven approach continuously monitors pumping, crystal temperature, alignment precision, and gain medium performance. When the data indicates a trend toward thermal instability, the system can automatically adjust the pump power or cooling setting or tweak the alignment using adaptive optics or motorized mirrors. The **compressor** is the final critical component that shortens the amplified pulse to its intended femtosecond duration. Small misalignments or fluctuations in the grating spacing or angles can lead to suboptimal compression, producing longer pulses and reducing peak power. A data-driven approach analyzes the output pulse width and phase in real time, comparing it against the target metrics. If the pulse begins to stretch, the machine learning model can correlate this with compressor misalignments or thermal shifts in the grating. Automated micro-adjustments to the grating positions or tilts can then be made to restore optimal pulse compression. This real-time feedback loop allows the compressor to maintain peak performance and adapt to changing environmental conditions, ensuring the pulse exits with minimal duration and maximum power.

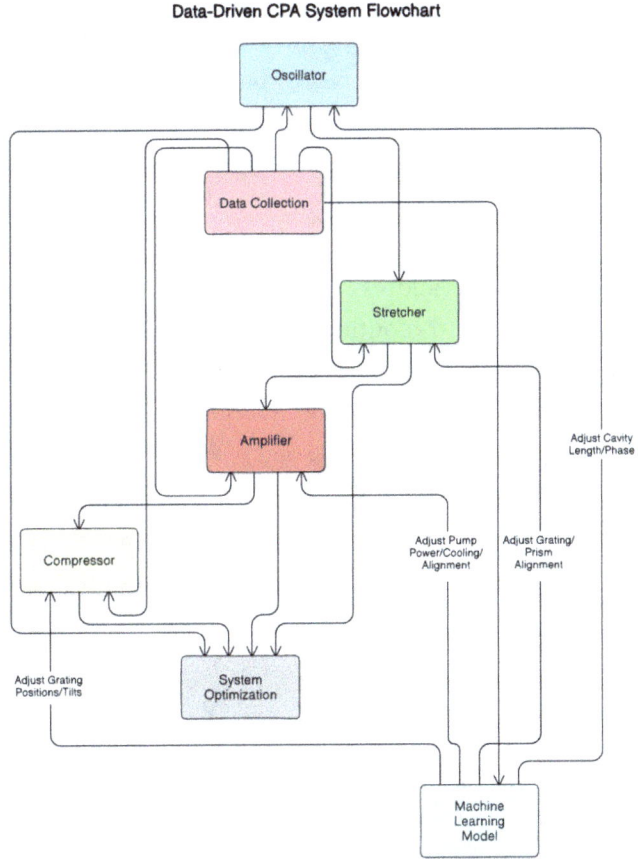

Overall, a data-driven approach to managing a Ti-sapphire-based CPA laser system provides a robust framework for stabilizing each component within the chain. The system can self-correct and optimize key parameters by leveraging machine learning models trained on extensive historical data combined with real-time monitoring and predictive diagnostics. This enhances pulse stability and quality and reduces the need for manual intervention, leading to lower maintenance costs and increased operational uptime.

The flowchart simplifies and summarizes the processes in a data-driven approach to optimizing and stabilizing a Ti-sapphire-based CPA laser system. It shows the core components of the system, which are the oscillator, stretcher, amplifier, and compressor, connected in sequence, with data collection gathering real-time data from each. This data is then analyzed by a machine learning model, which provides predictive insights and adjustments to maintain optimal performance across the system.

Each component has a specific feedback loop from the machine learning model: adjusting cavity length/phase for the Oscillator, grating or prism alignment for the Stretcher, pump power/cooling/alignment for the Amplifier, and grating positions/tilts for the Compressor. These feedback adjustments help address detected deviations from ideal performance, contributing to continuous system optimization. The design enables autonomous, adaptive operation of the CPA laser system, stabilizing key parameters for consistent pulse quality and overall system efficiency.

Future Trends in Laser Technology

The future of laser technology is set to undergo significant transformations driven by advancements in machine learning.

Predictions for Laser Technology Advancements

Exploration is ongoing to enhance laser performance through algorithms that optimize operating parameters in real-time. This integration allows for adaptive control of laser systems, improving precision and efficiency. By utilizing machine learning techniques, such as neural networks, scientists can analyze vast amounts of data generated during laser operations, identifying patterns that may not be apparent through traditional analysis methods. This predictive capability can lead to more reliable and robust laser systems, paving the way for new applications and innovations.

In medical applications, the combination of laser technology and machine learning is poised to revolutionize diagnostics and treatments. For instance, machine learning algorithms can more accurately analyze imaging data to pinpoint anomalies, allowing for more targeted laser therapy. Additionally, the ability to predict patient responses to laser treatments based on historical data could lead to personalized medicine approaches. As researchers continue to refine these algorithms, we can expect to see lasers that can adapt their operating conditions automatically based on real-time feedback from patient data, improving outcomes and minimizing risks during procedures.

Industrial applications of laser technology are also on the brink of transformation through machine learning integration. The precision of laser cutting and welding processes can be enhanced using real-time monitoring systems that leverage machine learning to predict and adjust for material variances and environmental conditions. These smart systems can reduce waste, improve product quality, and enhance safety in manufacturing environments. As industries increasingly adopt intelligent automation, the synergy between laser technology and machine learning will likely lead to increased productivity and reduced operational costs.

Furthermore, advancements in laser communication technologies, such as optical wireless communication, are expected to flourish with machine learning. Research is going on to enhance data transmission rates and reduce interference through adaptive modulation techniques informed by machine learning algorithms. This synergy could lead to the development of more efficient and reliable communication systems, particularly in applications requiring high bandwidth and low latency.

The future of laser technology will likely see a rise in the development of novel laser sources and materials driven by machine learning. Utilizing computational models to predict the behavior of new materials under laser excitation facilitates the discovery of advanced laser media with superior properties. This approach accelerates material discovery and enhances the design of lasers tailored for specific applications.

Predictions for Laser Technology Advancements, Laser Development, Laser Diagnostics, Implementation of Machine Learning

Advancements in laser technology, particularly in the context of laser development, diagnostics, and the integration of machine learning, are expected to bring significant improvements in precision, efficiency, and automation. With ongoing innovations in fiber, semiconductor, and solid-state lasers, Lasers will become more powerful and energy-efficient, leading to better beam quality and control. The miniaturization of laser systems will enable the development of compact and tunable lasers, essential for telecommunications, medical diagnostics, and quantum computing applications. New laser sources may also emerge from materials like exotic semiconductors or quantum dot lasers, broadening the range of applications and enhancing performance across different fields.

Laser diagnostics will evolve significantly, with real-time monitoring becoming a standard feature. Advanced sensors will provide instant feedback on laser operations, enabling automatic adjustments to maintain optimal performance and prevent malfunctions. Machine learning will play a key role in predictive maintenance, using operational data to predict when maintenance is required, thus minimizing downtime and extending the lifespan of laser systems. Lasers will become more stable and precise in fields like spectroscopy and imaging.

The integration of machine learning will transform laser systems into highly adaptive tools. Algorithms will dynamically adjust laser parameters in real-time, tailoring the system's performance to specific operating conditions and desired outcomes. These self-optimizing systems will analyze vast amounts of data to continuously improve efficiency, resulting in reduced waste and maximized output. In manufacturing, lasers used for cutting, welding, and engraving will become smarter, learning from previous operations and adapting settings for different materials to increase precision and automation. Machine learning will also be crucial for fault detection and correction, enabling lasers to detect deviations in performance and self-correct before these issues impact operations.

These advancements will greatly benefit emerging fields such as quantum computing and medical. In quantum computing, lasers are essential for controlling quantum bits (qubits), and machine learning will optimize laser performance for more stable and efficient quantum systems. In medical applications, machine learning will enhance the precision of lasers used in surgeries and diagnostics, reducing errors and improving outcomes in procedures such as laser-based surgeries or cancer detection. Environmental monitoring will also see improvements as laser-based remote sensing systems, enhanced by machine learning, will provide real-time insights into atmospheric conditions, pollution levels, and climate change indicators.

As lasers are increasingly integrated with other technologies, the collaboration between lasers and robotics will enable highly automated and precise systems in industries ranging from surgery to manufacturing. Lasers in augmented reality systems and 3D scanning will benefit from machine learning's ability to analyze large datasets, providing more accurate measurements and visual representations in architecture, medicine, and entertainment. The future of laser technology is poised for groundbreaking innovations, with machine learning driving a new era of smarter, more efficient, and autonomous laser systems.

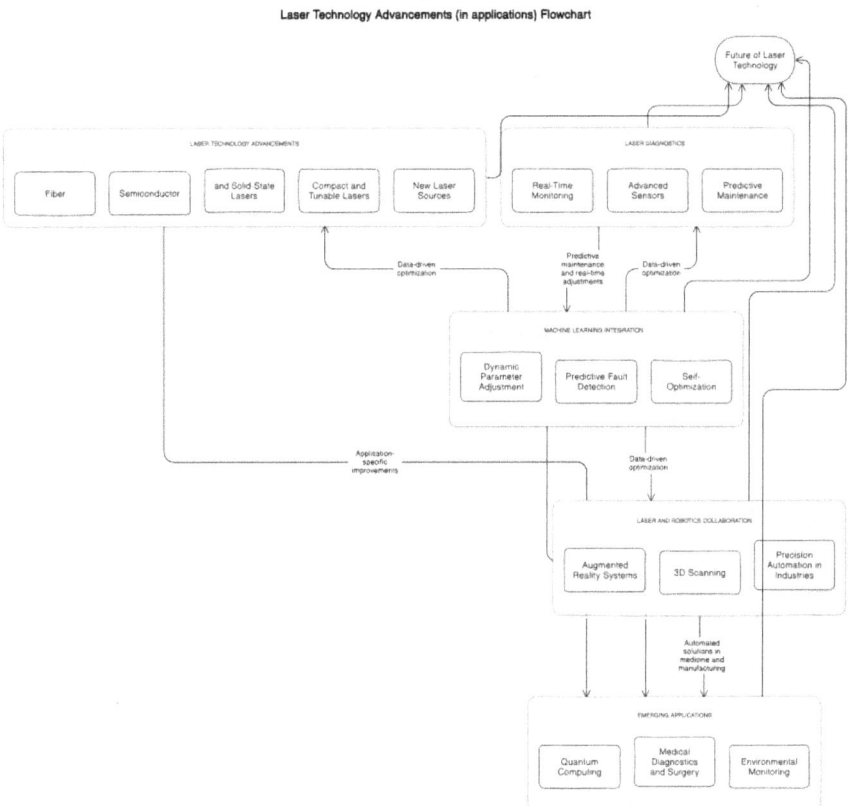

The block diagram diagrammatically explains the progression of laser technology advancements across various applications, highlighting key areas such as laser developments, diagnostics, machine learning integration, emerging applications, and robotics collaboration. Starting with laser technology advancement, innovations in fiber, semiconductor, solid-state lasers, compact tunable lasers, and new laser sources contribute to overall improvements. Laser diagnostics employs real-time monitoring, advanced sensors, and predictive maintenance to ensure continuous performance. Machine learning integration plays a key role, enabling dynamic parameter adjustment, predictive fault detection, and self-optimization, which feed back into diagnostics and technology advancements for data-driven optimization. The laser and robotics collaboration enables advancements in augmented reality systems, 3D scanning, and precision automation, which enhance applications in emerging applications such as quantum computing, medical diagnostics and surgery, and environmental monitoring. Finally, all these advancements converge toward the future of laser technology, representing a future of smarter, more efficient, and highly automated laser systems.

The Role of Machine Learning in Shaping Future Lasers

Machine learning is revolutionizing laser technology by improving how lasers are designed and performed. It can process large, complex datasets that traditional methods couldn't handle. By using algorithms to spot patterns and make predictions, machine learning boosts the efficiency of laser systems, unlocking innovations that once seemed impossible. This enhances current laser technologies and opens the door to new applications and possibilities.

One significant area where machine learning is profoundly impacting is the development of laser materials. The properties of laser gain media are critical for determining the efficiency and output of laser systems. Machine learning algorithms can analyze the relationships between material compositions, structural characteristics, and their corresponding laser performance metrics. This data-driven approach enables the identification of optimal material combinations that might not have been considered through conventional experimental methods. Consequently, it is better equipped to develop materials that enhance the performance and capabilities of lasers, including increased power output, improved stability, and broader wavelength coverage.

In addition to material development, machine learning is revolutionizing the optimization of laser manufacturing processes, streamlining production workflows, and minimizing defects by utilizing predictive modeling and real-time feedback mechanisms. Machine learning algorithms can analyze data from various stages of the manufacturing process, allowing for the identification of inefficiencies and anomalies. As a result, manufacturers can achieve higher precision and consistency in laser component fabrication, which is crucial for the reliability of laser systems in demanding applications like telecommunications, medical devices, and industrial cutting.

Machine learning also plays a vital role in the laser system's real-time monitoring and control. Advanced sensors combined with machine learning algorithms can provide continuous feedback on a laser's performance, allowing dynamic adjustments to respond to changing conditions. This capability is particularly important in applications where precision is important, such as laser surgery or materials processing. By ensuring that the laser operates at optimal parameters, machine learning increases the application's effectiveness and extends the laser's lifespan.

The synergy between machine learning and laser technology holds immense potential for future advancements. This could include the development of intelligent lasers capable of adapting to their environments or designing entirely new laser architectures based on data-driven insights. Harnessing the power of machine learning, the future of laser technology promises to be more versatile, efficient, and transformative than ever before, ultimately impacting a wide range of industries and applications.

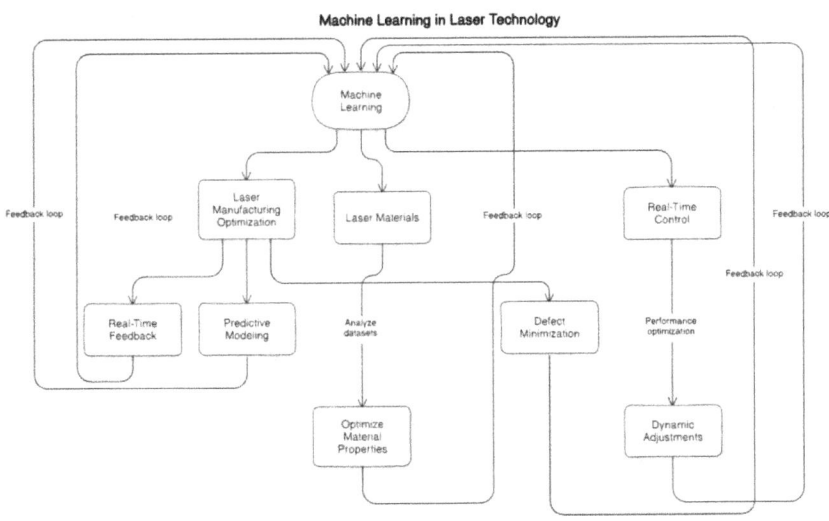

The role of machine learning is depicted in the diagram; it shows the integration of machine learning in laser technology, focusing on laser manufacturing optimization, real-time control, and material analysis. Machine learning drives optimization through real-time feedback and predictive modeling, which are used to enhance manufacturing efficiency and reduce defects. Laser materials are analyzed to optimize their properties, contributing to defect minimization. In parallel, real-time control systems rely on performance data to make dynamic adjustments for optimal laser operation. Feedback loops across all processes ensure continuous learning and improvement, creating a self-optimizing laser system.

Global Impacts of Laser Technology Development

The advancements in laser technology have far-reaching global impacts across various sectors, driven by innovations in machine learning and data-driven approaches. Modern laser technologies, including fiber, solid-state, chirped pulse amplification (CPA) systems, and disk lasers, are witnessing transformative improvements that enhance their precision, speed, and adaptability. As we explore new applications and efficiencies in laser systems, integrating machine learning algorithms enables lasers to perform complex tasks with unprecedented accuracy, transforming industries such as manufacturing, medical, and telecommunications. By harnessing data-driven insights, we can continuously monitor and optimize laser parameters in real-time, significantly improving outcomes, enhancing safety, and reducing waste. Machine learning and data-driven approaches have accelerated developments in power scalability and beam quality control in fiber lasers, making these lasers invaluable for manufacturing applications. Fiber lasers, which rely on optical fibers doped with rare earth elements as gain media, are known for their high efficiency and compact design. Machine learning algorithms analyze real-time data on beam quality and thermal effects, enabling precise adjustments to maintain stability, even at high power levels. This technology allows fiber lasers to tackle demanding applications in industries like automotive and aerospace, where precision in cutting, welding, and material processing is essential. Machine learning also enables predictive maintenance by monitoring operational data and anticipating issues before they disrupt production, further enhancing fiber laser's role in modern manufacturing. Solid-state lasers have been instrumental in various high-power and scientific applications. The integration of machine learning enables these lasers to operate more reliably and adaptively by dynamically adjusting parameters such as cooling, alignment, and power levels. Solid-state lasers are sensitive to thermal effects, which can distort the beam quality if not carefully managed. Data-driven insights allow real-time control over these variables, enhancing the stability and lifespan of the system. This ability to maintain consistent output in laser-based precision manufacturing has made solid-state lasers a trusted choice, as they can deliver high peak powers with minimal downtime.

The CPA laser system is another area where machine learning significantly impacts. CPA laser systems are essential for applications that require intense, ultrashort pulses, such as micromachining, material ablation, and scientific research. By analyzing real-time pulse width, dispersion, and alignment data, machine learning algorithms can make continuous adjustments to ensure that pulses remain within the optimal parameters. This data-driven approach prevents issues like self-focusing or beam distortion from high intensities, thereby protecting the CPA laser system from potential damage. Machine learning also helps manage the complex interplay between components in a CPA laser system, such as the oscillator, stretcher, amplifier, and compressor, enabling enhanced performance and greater pulse consistency.

Disk lasers, characterized by their thin, disk-shaped gain medium, offer high efficiency and excellent heat dissipation, making them suitable for high-power scientific applications and industrial applications like heavy-duty cutting and welding. In these systems, machine learning algorithms are vital in optimizing cooling and alignment to mitigate thermal lensing, which can distort the beam as power levels rise. With a data-driven approach, disk lasers can dynamically adjust to different materials and environmental conditions, enhancing flexibility and operational efficiency. This adaptability is especially valuable in industries that require prolonged, high-power operation, as it reduces the need for manual adjustments and ensures continuous, precise performance. Furthermore, machine learning enables disk lasers to adapt to various manufacturing demands, streamlining operations and improving productivity.

In manufacturing, the evolution of laser cutting, welding, and engraving technologies is revolutionizing production processes. Machine learning models analyze operational data to identify inefficiencies and predict maintenance needs, minimizing downtime and maximizing throughput. Predictive maintenance enabled by machine learning reduces operational costs and enhances productivity, making laser technology a cornerstone of modern manufacturing. As companies adopt these advanced systems, there is a growing demand for skilled professionals who can leverage both laser technology and machine learning, shaping the workforce of the future. The medical field is also undergoing transformative changes due to laser technology. From laser-based surgical procedures to advanced diagnostic tools, the precision and customization offered by lasers are significantly improving patient outcomes. Machine learning algorithms aid in analyzing vast amounts of medical data, facilitating the development of personalized treatment plans incorporating laser therapies. For example, in laser-based cancer treatments, machine learning can help adjust the laser's intensity and targeting parameters based on individual patient characteristics, increasing effectiveness while reducing potential side effects. The combination of data-driven insights and real-time laser adjustments enhances medical procedure's safety, accuracy, and effectiveness, illustrating the powerful potential of integrating machine learning with laser technology in medicine. In telecommunications, advancements in laser technologies, particularly in fiber-optic communications, have been crucial for high-speed data transmission. The precision of laser light enables it to carry information over long distances with minimal loss, making it essential for modern communication networks. Machine learning enhances this technology by optimizing signal processing and error correction, leading to more reliable and faster communication networks. By continuously analyzing transmission data, machine learning algorithms can dynamically adjust signal parameters to maintain optimal performance, even in the face of environmental fluctuations or increased data traffic.

The global impacts of laser technology developments extend beyond specific industries, fostering interdisciplinary collaboration and driving economic growth. Ultimately, the fusion of laser technology and machine learning enhances existing processes and paves the way for future advancements, reshaping the possibilities of laser applications in unprecedented ways and contributing to a smarter, more connected, and more efficient world.

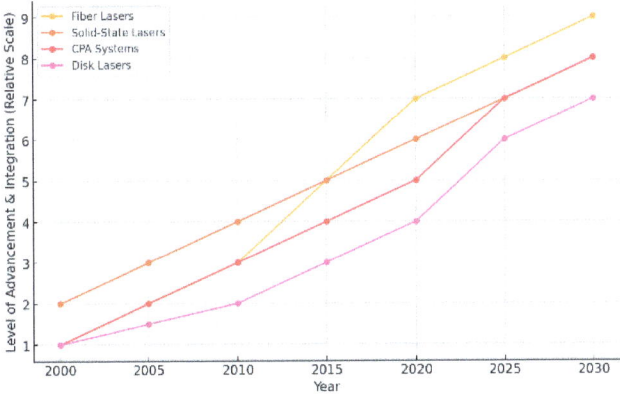

The predictive graph illustrates the timeline of advancements in laser technology integration with data-driven and machine learning approaches from 2000 to 2030. The graph tracks the progression of some laser types, fiber lasers, solid-state lasers, CPA systems, and disk lasers, showing increasing levels of development and integration over the years. Fiber lasers demonstrate rapid growth in their use of machine learning for real-time adjustments and predictive maintenance, making them highly adaptable in manufacturing applications. Solid-state lasers also see steady thermal management and stability improvements, particularly for high-power applications. CPA laser systems, essential for ultrashort pulse generation, benefit significantly from data-driven optimization for pulse stability and consistency. Disk lasers show slower initial growth but rise steadily as machine learning enhances their cooling and alignment adjustments for high-power industrial use. This graph highlights the expected advancements and growing role of machine learning in enhancing laser technology's precision, adaptability, and operational efficiency over time.

Do We Need Coding for Machine Learning Applications in Lasers?

As machine learning becomes increasingly accessible, a pressing question arises: do you need to learn coding to use machine learning in laser applications effectively? Traditionally, coding was a cornerstone for any work involving machine learning in lasers. Scientists and engineers needed to develop algorithms to analyze beam profiles, preprocess experimental data, and implement mathematical models tailored to their systems. Programming languages like Python, R, and MATLAB, complemented by libraries such as TensorFlow, PyTorch, and Scikit-learn, provided the tools necessary for these tasks. Coding enabled unparalleled control, allowing us to design and fine-tune models for highly specific challenges, such as optimizing beam quality, diagnosing laser alignment errors, and predicting equipment failures. For example, a laser scientist working to stabilize ultrashort pulses in a chirped pulse amplification system might develop machine learning models to predict pulse distortions under varying environmental conditions. Similarly, data scientists studying thermal effects in high-power lasers could create custom scripts to preprocess temperature data and uncover patterns leading to thermal lensing. These efforts, however, came with a steep learning curve, requiring a combination of programming proficiency and deep knowledge of physics and mathematics.

New no-code and low-code platforms have made machine learning much easier to use by removing many of the challenges that used to make it difficult to get started. Tools like Google AutoML, IBM Watson Studio, and Microsoft Azure Machine Learning automate critical tasks such as data preprocessing, feature selection, and model training, enabling professionals to implement machine learning without extensive coding experience. These platforms allow for addressing complex problems without delving into the underlying code. For instance, an optics engineer might classify laser beam profiles by symmetry and intensity using a no-code tool by simply uploading a dataset. The platform handles the technical details, producing a classification model in minutes. Likewise, a maintenance manager could use historical data on power output and beam stability to predict when a high-power laser will need servicing. This accessibility has accelerated the adoption of machine learning in laser technology.

Despite their simplicity, no-code platforms are not without limitations. Their reliance on pre-configured workflows can restrict their usefulness for advanced or highly customized applications. For example, designing machine learning models for real-time beam shaping or optimizing multi-pass amplification often requires more flexibility than these platforms provide. No-code tools also tend to function as "black boxes," obscuring the logic behind their decisions. This lack of transparency can be problematic in applications where understanding model behavior is crucial, such as stabilizing femtosecond pulse durations or designing custom optical coherence tomography systems.

Furthermore, no-code solutions may struggle to scale effectively for large, complex laser setups. These limitations, however, do not diminish their value for routine tasks. They enable faster prototyping, democratizing access to machine learning, and allowing laser professionals to focus on interpreting results and making informed decisions. Coding remains indispensable in scenarios requiring advanced customization, scalability, or integration, especially in laser research and development. For example, optimizing beam quality in high-power lasers often demands custom scripts to analyze the intricate relationships between optical components, environmental conditions, and output energy. Large-scale systems, such as those used in industrial laser cutting or additive manufacturing, require coded solutions to manage real-time feedback and optimize resource usage. Coding helps to explore new frontiers, such as applying deep learning to study nonlinear optical phenomena or developing reinforcement learning algorithms for adaptive laser control. For example, predicting thermal lensing in solid-state lasers might require creating a custom machine learning system that uses specific knowledge about how heat spreads and how the laser beam behaves. Beyond research, many laser systems need integration with broader operational frameworks, such as robotic automation or IoT-enabled diagnostics, where programming is essential.

The future of machine learning in lasers likely lies in a hybrid approach that combines the strengths of coding and no-code platforms. Platforms like Microsoft Azure Machine Learning and Google AutoML allow users to begin with simplified interfaces and transition to code for advanced customization. This flexibility caters to both beginners and experienced users. For instance, a laser scientist might use AutoML to quickly identify a promising model for predicting alignment errors and then refine it with Python scripts to account for specific experimental nuances. Similarly, no-code platforms could streamline routine tasks like initial data cleaning and feature selection, while coding facilitates in-depth analysis of phenomena such as beam coherence under varying conditions. By bridging the gap between accessibility and customization, this hybrid approach ensures that machine learning remains a powerful tool for rapid application and detailed exploration.

Whether coding is necessary for machine learning in lasers ultimately depends on the goals and complexity of the application. No-code platforms offer a straightforward and efficient solution for many practical scenarios, such as monitoring laser system health or classifying beam profiles. However, coding remains irreplaceable for tasks requiring deeper insights, such as real-time stabilization of multi-pass amplifiers or modeling interactions in ultrafast optics. The choice between coding and no-code solutions is not binary but contextual, reflecting the specific demands of the task. Today, the availability of advanced tools and platforms ensures that machine learning is accessible to laser professionals of all expertise levels, marking an exciting era of innovation in laser technology. Whether you choose to code or not, the opportunities to leverage machine learning in lasers are vast and growing. Suppose we extend our discussion in the form of a diagram. In that case, it will illustrate the decision-making process and workflows in applying machine learning to laser technology, emphasizing the balance between traditional coding and no-code platforms. It begins by introducing the central question of whether coding is necessary for machine learning in lasers.

The diagram splits into two parallel paths: the traditional coding approach, which involves algorithm development, data preprocessing, and addressing challenges like a steep learning curve, and the no-code platform approach, which highlights simplified workflows using tools for accessibility. These paths converge into a hybrid approach, demonstrating the integration of both methods to leverage each other's strengths.

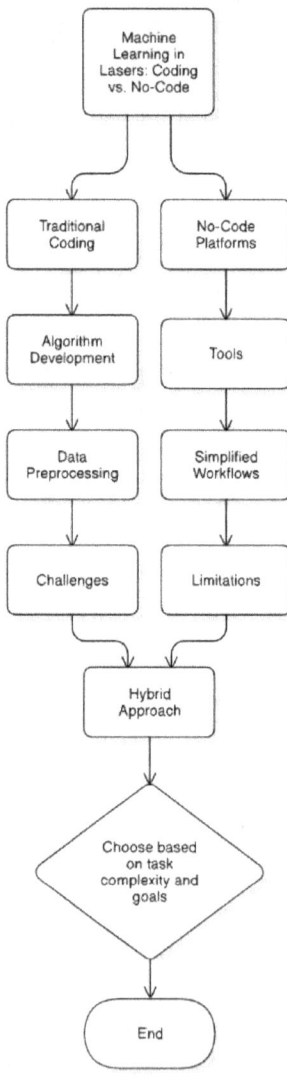

The process concludes with a decision based on the complexity and goals of the task, highlighting that the choice between coding and no-code approaches is contextual. This diagram visually encapsulates the narrative of balancing accessibility with customization in laser applications using machine learning.

Challenges, Opportunities, Conclusion, and Outlook

The convergence of laser technology and machine learning represents a transformative shift with far-reaching implications across industries.

Challenges and Opportunities

The intersection of laser technology and machine learning presents a landscape rich with both challenges and opportunities. As research progresses, the complexities of integrating these two fields become evident. One major challenge is the need for specialized knowledge spanning both domains. Must be well-versed not only in the principles of laser technology but also in the algorithms and methodologies of machine learning. This dual expertise is crucial for developing innovative applications that leverage the strengths of both fields, such as improving laser manufacturing processes or enhancing precision in laser-based diagnostics. Another significant hurdle is the data acquisition and processing requirements associated with machine learning. High-quality, diverse datasets are essential for training effective machine learning models. In laser technology, gathering such data can be challenging due to the variability in laser systems and their operating conditions. Moreover, integrating sensors and imaging technologies to capture real-time data adds another layer of complexity, necessitating interdisciplinary collaboration among scientists and engineers.

Despite these challenges, significant opportunities arise by applying machine learning techniques to enhance laser technology. For instance, machine learning algorithms can optimize laser parameters in real-time, significantly improving efficiency and outcomes in industrial applications. Predictive modeling can also be employed to foresee equipment failures or maintenance needs, thus reducing downtime and associated costs. Furthermore, machine learning can facilitate the development of adaptive laser systems that can adjust their performance based on environmental changes or specific material characteristics, leading to broader applications across various industries.

The educational landscape also stands to benefit from this convergence. As laser technology and machine learning continue to evolve, there is a growing need for curricula that encompass both fields. Institutions can cultivate a new generation of engineers and researchers adept in laser systems and machine learning techniques. By fostering interdisciplinary programs, universities can better prepare students for careers that require a fusion of these technologies, ensuring they possess the necessary skill sets to tackle future challenges and leverage emerging opportunities. While integrating laser technology and machine learning poses several challenges, it simultaneously opens doors to innovative solutions and advancements. Researchers and practitioners are encouraged to embrace these complexities as they work towards harnessing the full potential of both fields. By addressing the knowledge gaps, enhancing data collection methods, and fostering interdisciplinary education, stakeholders can not only overcome existing obstacles but also pave the way for groundbreaking developments that will shape the future of laser technology.

Summary of Key Insights

Modern laser systems, including fiber lasers, solid-state lasers, CPA laser systems, and disk lasers, are progressing rapidly with the integration of machine learning, bringing unprecedented levels of precision, efficiency, and adaptability. By analyzing operational data in real-time, machine learning algorithms can optimize parameters such as intensity, power, focusing, and duration, which enhances output quality, reduces operational time, and minimizes resource use. These capabilities are crucial for manufacturing, where precise and efficient processes are essential for productivity and cost savings.

Machine learning also plays a pivotal role in predictive maintenance for laser systems. By employing predictive analytics, laser systems can anticipate maintenance needs or potential failures, allowing proactive actions that prevent costly downtime. This is particularly important for high-demand applications, such as fiber lasers in automotive and aerospace industries or solid-state lasers in scientific and medical fields. Real-time data analysis can detect patterns that signal wear or malfunctions, ensuring that laser systems remain reliable and operationally efficient. The ability to predict maintenance needs extends the lifespan of laser equipment, enhancing productivity and reducing the risk of unexpected disruptions.

Moreover, the adaptability of machine learning enhances the versatility of laser applications across various domains. Fiber lasers, with their high efficiency and power scalability, benefit significantly from machine learning, which enables real-time adjustments in beam quality and power based on the material or task requirements. In medical applications, machine learning allows solid-state and disk lasers to adjust parameters dynamically, offering customized treatments in surgeries and diagnostics, improving patient outcomes, and reducing recovery times. This adaptability opens new opportunities in fields previously limited by conventional laser technologies, paving the way for innovations in areas such as telecommunications, where laser light precision in fiber-optic communication is enhanced by machine learning for signal optimization and error correction.

Furthermore, machine learning facilitates data visualization and analysis, simplifying complex data produced by laser diagnostics and experimental setups. High-dimensional data from laser experiments, such as in CPA systems, can be challenging to interpret; however, machine learning algorithms can uncover hidden trends and provide actionable insights. For instance, machine learning can help optimize the pulse width, dispersion, and alignment in CPA laser systems to prevent issues like self-focusing or beam distortion. Enhanced data interpretation and visualization empower researchers and engineers to make more informed decisions, accelerating the development and application of laser technologies across industries.

The collaboration between laser technology and machine learning also emphasizes the need for interdisciplinary approaches. Physicists, engineers, data scientists, and industry experts must work together to fully harness this integration's potential. This collaboration will drive innovation and foster educational programs that prepare future professionals to navigate and contribute to this rapidly evolving landscape. The future of laser technology, empowered by machine learning, is dynamic and transformative, holding immense promise for reshaping numerous sectors globally.

Future Directions for Research

The intersection of laser technology and machine learning provides fertile ground for future research, particularly in refining the integration of machine learning algorithms into various laser systems. Focus on advancing the predictive models that analyze real-time data to facilitate adaptive control of laser parameters such as power, wavelength, and beam focusing. This approach will enhance precision in laser applications, enabling advancements in material processing, biomedical applications, and scientific research.

A promising direction involves developing hybrid systems that combine traditional laser technologies with machine learning frameworks. For example, integrating neural networks with laser diagnostics could improve the characterization of laser-material interactions, informing the design of more effective laser systems. Future research could focus on building robust datasets encompassing a wide range of materials and operating conditions, allowing machine learning algorithms to learn from diverse scenarios and enhance their predictive capabilities. This is especially relevant for applications in fields such as environmental monitoring and industrial manufacturing, where laser technologies face a variety of material and environmental variables.

Generative models, such as Generative Adversarial Networks (GANs), also present an exciting research avenue. GANs can simulate and optimize laser beam profiles for specific applications, generating synthetic data that mimics real-world conditions. This enables researchers to explore optimal configurations for different laser applications, including cutting, welding, and medical treatments. Training machine learning models with generated data can accelerate the development of tailored laser technologies, offering more precise and customized solutions for industrial and medical needs.

Collaborative research across disciplines will advance machine learning and laser technology integration. Partnerships between physicists, engineers, computer scientists, and industry practitioners can leverage diverse expertise to address complex challenges in laser applications.

Final Thoughts on the Convergence of Laser Technology and Machine Learning

The convergence of laser technology and machine learning marks a major milestone, enhancing precision, efficiency, and adaptability across various applications. Machine learning algorithms provide previously unattainable insights by analyzing extensive datasets generated during laser operations. These insights optimize critical parameters such as power, beam profiling, pulse duration, and focusing techniques, improving laser output quality and reducing waste.

Laser systems with machine learning capabilities adapt dynamically to changing material properties and environmental conditions, especially in manufacturing. This enables lasers to learn from previous cutting or engraving tasks, adjusting in real-time to ensure consistent quality and precision. As a result, manufacturers can achieve higher throughput and maintain stringent quality standards, broadening the scope of laser applications across various industrial sectors.

Integrating laser technology and machine learning in the medical field has also unlocked new avenues for diagnostics and treatment. Machine learning algorithms can analyze medical images produced by laser systems, such as those used in ophthalmology or dermatology, assisting clinicians in making more accurate diagnoses. By recognizing patterns and anomalies, machine learning enhances the capabilities of laser-based procedures, resulting in better patient outcomes. Predictive analytics further personalizes treatment plans, tailoring laser-based interventions to individual patient needs and improving the effectiveness of medical procedures.

The educational sector also embraces this convergence, with curricula increasingly incorporating machine learning and laser technology. By understanding the impact of machine learning on laser applications, students and researchers gain the skills necessary to drive future advancements. Hands-on projects that involve real-time data analysis and machine learning applications in laser experiments offer a comprehensive learning experience, preparing students to innovate at the forefront of these fields.

In conclusion, the synergy between laser technology and machine learning promises transformative changes across multiple sectors. This integration enhances laser systems capabilities, promotes innovation, and improves efficiency in various applications.

References and Further Reading

Books

1. Silfvast, W. T. (2004). *Laser Fundamentals*. Cambridge University Press.
2. Svelto, O. (2010). *Principles of Lasers* (5th ed.). Springer.
3. Milonni, P. W., & Eberly, J. H. (2010). *Laser Physics*. Wiley.
4. Hitz, C. B., Ewing, J. J., & Hecht, J. (2012). *Introduction to Laser Technology* (4th ed.). Wiley.
5. Ebersbach, D., & Gottschalk, F. (Eds.). (2013). *Laser Technology, Applications and Future Prospects (Lasers and Electro-Optics Research and Technology: Physics Research and Technology)*. Nova Novinka.
6. Titterton, D. H. (2015). *High-Power Laser Technology and Systems*. SPIE Press.
7. Shulika, O., & Sukhoivanov, I. (Eds.). (2015). *Advanced lasers: Laser physics and technology for applied and fundamental science* (Vol. 193). Springer.
8. Nolte, S., Schrempel, F., & Dausinger, F. (Eds.). (2015). *Ultrashort Pulse Laser Technology: Laser Sources and Applications* (Springer Series in Optical Sciences, Vol. 195). Springer.
9. Goodfellow, I., Bengio, Y., & Courville, A. (2016). *Deep Learning*. MIT Press.
10. Prabhu, S. R. J. S. (2018). *Photonics and Machine Learning*. Springer
11. Géron, A. (2019). *Hands-On Machine Learning with Scikit-Learn, Keras, and TensorFlow* (2nd ed.). O'Reilly Media
12. Brunton, S. L., & Kutz, J. N. (2019). *Data-Driven Science and Engineering: Machine Learning, Dynamical Systems, and Control*. Cambridge University Press.
13. Parker, D. A. A. (2019). *Machine Learning for Laser Systems*. Springer.
14. Alpaydin, E. (2020). *Introduction to Machine Learning* (4th ed.). MIT Press.
15. Lee, K. E. M. (2020). *Introduction to Machine Learning for Photonics and Laser Applications*. CRC Press.

16. Ernst, F., & Schweikard, A. (2020). *Fundamentals of machine learning*. UTB GmbH.
17. McClarren, R. G. (2021). *Machine learning for engineers: Using data to solve problems for physical systems* (1st ed.). Springer.
18. Sharma, O. D. (2021). *Machine Learning for Absolute Beginners: A Plain English Introduction* (3rd ed.). Independently published.
19. Kistenev, Y. V., Borisov, A. V., & Vrazhnov, D. A. (2021). *Medical Applications of Laser Molecular Imaging and Machine Learning* (1st ed.). SPIE--The International Society for Optical Engineering.
20. Lau, A. P. T., & Khan, F. N. (Eds.). (2022). *Machine Learning for Future Fiber-Optic Communication Systems* (1st ed.). Academic Press.
21. Krüger, J., & Bonse, J. (Eds.). (2023). *Advanced pulse laser machining technology.* MDPI
22. Lee, T.-F., & Liu, C.-H. (2024). *Advanced laser techniques and data-driven strategies for skin treatment: Bridging technology and clinical practice.* Eliva Press.

Articles

1. A. M. Turing, I.—Computing Machinery and Intelligence, *Mind*, Volume LIX, 236, 430-460 (1950). https://doi.org/10.1093/mind/LIX.236.433
2. Tercan, H., Khawli, T.A., Eppelt, U. *et al.* Improving the laser cutting process design by machine learning techniques. *Prod. Eng. Res. Devel.* 11, 195–203 (2017). https://doi.org/10.1007/s11740-017-0718-7
3. J. Fang, A. Swain, R. Unni, Prof. Y. Zheng, "Decoding Optical Data with Machine Learning," Laser Photonics Rev. 15, 2000422 (2021). https://doi.org/10.1002/lpor.202000422
4. Mills, B., & Grant-Jacob, J. A. Lasers that learn: The interface of laser machining and machine learning. *IET Optoelectronics*, *15*(5), 207-224 (2021). https://doi.org/10.1049/ote2.12039
5. Sergey M. Kobtsev, "Perspective paper: Can machine learning become a universal method of laser photonics?," *Optical Fiber*

Technology, 65, 102626 (2021).
https://doi.org/10.1016/j.yofte.2021.102626
6. Lucas R. Hofer, Milan Krstajić, and Robert P. Smith, "Measuring laser beams with a neural network," *Appl. Opt.* 61, 1924-1929 (2022). https://doi.org/10.1364/AO.443531
7. Kuprikov, E., Kokhanovskiy, A., Serebrennikov, K. *et al.* Deep reinforcement learning for self-tuning laser source of dissipative solitons. *Sci Rep* 12, 7185 (2022). https://doi.org/10.1038/s41598-022-11274-w
8. Vives J, Palací J, "Artificial Intelligence and 3D Scanning Laser Combination for Supervision and Fault Diagnostics," *Sensors (Basel), MDPI*, 22 (19), 7649 (2022). https://doi.org/10.3390/s22197649
9. Döpp A, Eberle C, Howard S, Irshad F, Lin J, Streeter M. Data-driven science and machine learning methods in laser–plasma physics. *High Power Laser Science and Engineering*.11:e55 (2023). https://doi.org/10.1017/hpl.2023.47
10. Aldoseri, A., Al-Khalifa, K.N., Hamouda, A.M., "Re-Thinking Data Strategy and Integration for Artificial Intelligence: Concepts, Opportunities, and Challenges," *Appl. Sci.* 13, 7082 (2023). https://doi.org/10.3390/app13127082
11. Ma, Q., Yu, H. Artificial Intelligence-Enabled Mode-Locked Fiber Laser: A Review. *Nanomanuf Metrol* 6, 36 (2023). https://doi.org/10.1007/s41871-023-00216-3
12. Zhiwei Fang, Guoqing Pu, Yongxin Xu, Weisheng Hu, and Lilin Yi, "Data-driven inverse design of mode-locked fiber lasers," *Opt. Express* 31, 41794-41803 (2023). https://doi.org/10.1364/OE.503958
13. Zhao, H., Xu, D., Wu, Z., Sun, L., Yuan, G., & Wang, Z., "High-LinearFrequency-Swept Lasers with Data-Driven Control," *Photonics*, 10 (9), 1056 (2023). https://doi.org/10.3390/photonics10091056
14. Daniel Mostowski, Krzysztof Jakubczak, Piotr Garbat, "Automated laser beam characterization using artificial

intelligence (AI) for the predictive maintenance of lasers," *Optics & Laser Technology*, 177, 111087 (2024). https://doi.org/10.1016/j.optlastec.2024.111087
15. Ruei-Yu Huang, Jun-Qi Lu, Chung-Wei Cheng, Mi-Ching Tsai, An-Chen Lee, "Multi-data-driven model-based control to improve the accuracy of overhang structures in laser powder bed fusion," *Optics & Laser Technology* 171, 110398 (2024). https://doi.org/10.1016/j.optlastec.2023.110398
16. V. Gaciu, I. Dăncuș, B. Diaconescu, D. G. Ghiță, E. Slușanschi, C. M. Ticoș, "Classification of laser beam profiles using machine learning at the ELI-NP high power laser system. *AIP Advances*, 14 (4), 045114 (2024). https://doi.org/10.1063/5.0195174
17. Santos, E.P., Silva, R.F., Maciel, C.V.T. *et al.* Investigation of Random Laser in the Machine Learning Approach. *Braz J Phys* **54**, 70 (2024). https://doi.org/10.1007/s13538-024-01452-8

Appendix

A. Glossary of Key Terms

This glossary includes definitions for some essential terms in laser technology and machine learning, helping readers understand essential concepts in laser technology and machine learning.

Chirped Pulse Amplification (CPA): A process that temporarily stretches and amplifies laser pulses, allowing for high-power output without damaging optical components. CPA is widely used in applications requiring ultrafast, high-energy pulses.

Femtosecond Lasers: Lasers that produce ultrashort pulses lasting only femtoseconds (10^{-15} seconds), enabling precise control in fields like micromachining and ophthalmology.

Fiber Lasers: High-efficiency lasers that use optical fibers doped with rare-earth elements as the gain medium. Known for their compactness and beam quality, fiber lasers are commonly used in industrial applications like cutting and welding.

Laser Cutting: A process that uses focused laser beams to cut materials with high precision, commonly applied in manufacturing industries to achieve intricate designs and clean edges.

Laser Welding: A joining process that uses concentrated laser energy to fuse materials. Laser welding is ideal for applications requiring minimal thermal distortion, such as automotive and aerospace manufacturing.

Medical Lasers: Tailored for medical procedures, including surgery, dermatology, and ophthalmology. Medical lasers offer precise, minimally invasive treatment options with applications in corrective eye surgery, skin treatments, and cancer therapy.

Machine Learning: A field within artificial intelligence that enables systems to improve performance through data analysis and pattern recognition without explicit programming.

Supervised Learning: A machine learning approach where models are trained on labeled data, often used in predictive diagnostics and process optimization in laser systems.

Unsupervised Learning: A learning technique that finds patterns or clusters within unlabeled data, useful for anomaly detection and analysis in laser diagnostics.

Reinforcement Learning: A machine learning method where models learn through interaction with an environment, adjusting parameters based on rewards or penalties, used in adaptive control for real-time adjustments in laser processing.

Ti-Sapphire Laser: A widely used tunable laser with titanium-doped sapphire as the gain medium, known for generating ultrafast pulses and supporting high-power applications.

LiDAR (Light Detection and Ranging): A remote sensing technology that uses laser light to measure distances and create high-resolution maps, extensively used in autonomous vehicles and environmental monitoring.

B. Summary of Technical Diagrams and Schematics

This section provides an overview of key technical diagrams and schematics discussed throughout the book, serving as a quick reference for readers to locate visual aids related to important concepts in laser technology and machine learning applications.

Chirped Pulse Amplification (CPA) System Diagram

Location: Chapter on Advanced Laser Technologies

Description: Illustrates stages in a CPA system, highlighting pulse stretching, amplification, and compression with machine learning integration points.

Machine Learning Workflow for Laser Optimization

Location: Chapter on Machine Learning Techniques in Laser Applications

Description: Depicts the data flow and model training process in a machine learning pipeline for laser optimization, from data acquisition to real-time feedback.

Laser Cutting and Welding Optimization Diagram

Location: Chapter on Case Studies Convergence of Machine Learning And Laser Technology

Description: Shows the use of machine learning to dynamically adjust laser parameters in cutting and welding processes for industrial precision.

Medical Laser System with Machine Learning Feedback

Location: Chapter on Case Studies Convergence of Machine Learning And Laser Technology

Description: Details a feedback loop where machine learning optimizes laser settings based on patient data for improved safety in medical treatments.

LiDAR Data Processing Flow

Location: Chapter on Machine Learning Techniques in Laser Applications

Description: Demonstrates LiDAR data processing for object detection and environmental mapping using machine learning algorithms.

Predictive Maintenance in Automated Laser Systems

Location: Chapter on Machine Learning Techniques in Laser Applications

Description: Outlines the workflow for predictive maintenance, where machine learning detects potential issues in laser equipment to minimize downtime.

Machine Learning in Ti-Sapphire CPA Systems

Location: Chapter on Case Studies Convergence of Machine Learning And Laser Technology

Description: Visualizes how machine learning models adjust parameters in a Ti-Sapphire CPA system for real-time optimization.

C. Overview of Machine Learning Algorithms

This section overviews the machine learning models and algorithms for optimizing laser technology applications. These methods allow for better predictive accuracy, adaptive control, and data-driven decision-making in laser processes.

Predictive Modeling: Predictive modeling techniques, including regression models and decision trees, frequently forecast outcomes based on historical data. In laser technology, they can predict the optimal parameters for cutting, welding, or other processes. For example, linear regression effectively predicts the relationship between laser power and cutting speed, while more complex models, like random forests, provide enhanced prediction accuracy by averaging the results from multiple decision trees.

Neural Networks: Neural networks, particularly deep learning models, are useful for recognizing complex patterns in large datasets. In laser diagnostics, neural networks can analyze sensor data to detect subtle shifts in system performance or identify faults before they become significant issues. Convolutional Neural Networks (CNNs), a subtype of neural networks, are especially useful in image-based laser applications, such as laser-based imaging or LiDAR, enabling high accuracy in object detection, classification, and segmentation.

Support Vector Machines (SVMs): SVMs are powerful for classification tasks and are particularly useful when data has clear boundaries. In laser manufacturing, SVMs can classify different materials based on their response to laser irradiation, optimizing parameters for each material type. Additionally, SVMs are applied in quality control, where they can differentiate between acceptable and defective products based on laser scan data.

Reinforcement Learning (RL): RL is a machine learning technique where an agent learns by interacting with its environment and receiving feedback through rewards or penalties. In laser technology, RL algorithms are used in adaptive control systems, such as dynamically adjusting laser parameters in response to environmental changes during laser welding or cutting. This allows the system to optimize performance in real-time, ensuring consistent quality under varying conditions.

Clustering Algorithms: Clustering methods, such as K-means and hierarchical clustering, are employed to group data points with similar characteristics in unsupervised learning. In laser diagnostics, clustering helps identify patterns in operational data, such as grouping similar laser parameter settings that yield optimal performance. It's also useful for identifying anomalies, allowing for early detection of irregularities in laser system behavior.

Principal Component Analysis (PCA): PCA is a dimensionality reduction technique that simplifies complex datasets by transforming them into uncorrelated components. In laser applications, PCA can reduce the dimensionality of data generated from multi-sensor systems, such as simultaneously monitoring various laser parameters. By focusing on the most critical components, PCA aids in extracting meaningful insights while improving computational efficiency.

Gaussian Mixture Models (GMMs): GMMs are probabilistic models representing data as a mixture of several Gaussian distributions. They are beneficial in laser diagnostics, where they can model the distribution of parameters like beam intensity and pulse duration. GMMs are also used in anomaly detection, where they help identify deviations from normal operating conditions, supporting predictive maintenance efforts.

Bayesian Networks: These probabilistic models use Bayes' theorem to predict the likelihood of different outcomes. Bayesian networks are helpful in laser systems where uncertainty exists, as they can model complex dependencies among variables, such as the relationship between laser power, beam quality, and environmental factors. By incorporating prior knowledge, Bayesian networks allow for more robust decision-making in laser diagnostics and predictive maintenance.

Genetic Algorithms (GAs): GAs are optimization algorithms inspired by natural selection. They are used to optimize multi-objective problems in laser design, such as balancing beam quality and power efficiency. In laser manufacturing, GAs can optimize pulse duration and intensity by iteratively selecting the best-performing configurations, enhancing overall system performance.

Ensemble Methods: Ensemble learning methods, such as Random Forests and Gradient Boosting, combine multiple machine learning models to improve prediction accuracy. In laser diagnostics, ensemble methods help integrate data from different sensors to comprehensively understand system performance. For instance, Gradient Boosting can predict laser output stability by aggregating predictions from multiple weak learners, making the system more resilient to noise and errors.

Anomaly Detection: Algorithms specifically designed for anomaly detection, such as Isolation Forests and Local Outlier Factor (LOF), are critical in maintaining laser system reliability. These algorithms detect unusual patterns in operational data, indicating potential issues such as misalignment or wear in laser components. Anomaly detection is essential for predictive maintenance, where early detection of irregularities helps prevent costly downtime and repairs.

ACKNOWLEDGMENTS

This book would not have been possible without the grace and guidance of Allah, the All-Mighty, to whom I am deeply grateful for His blessings and inspiration throughout this journey.

My heartfelt appreciation goes to my beloved wife, whose encouragement and steadfast support have been my greatest source of strength. Thank you for believing in me, for your patience during the long work hours, and for making sacrifices that allowed me the time and focus needed to complete this project.

To my son, whose curiosity and energy inspire me every day, thank you for bringing so much joy into my life. Your presence is a reminder of why I strive to be my best.

I would also like to thank my colleagues, students, and friends, who generously shared their insights, guidance, and feedback. Each of you has enriched my work in invaluable ways, and I am fortunate to have had your support.

Finally, I wish to thank my readers. May this book offer knowledge, insight, or inspiration, just as it has been my journey to create it.

INDEX

1950, 123

A

ablation, 13, 31, 40, 63, 105
abundance, 18
accelerate, 118
accuracy, 18, 22, 28, 34, 35, 46, 48, 57, 59, 71, 73, 87, 90, 91, 104, 106, 125, 128, 129, 130
acquisition, 48, 65, 66, 86, 115
across, 36, 51, 53, 99, 104, 116, 117, 119, 120
actions, 30, 34, 116
adapt, 53, 65, 98, 105, 119
adaptability, 2, 22, 24, 38, 40, 48, 51, 53, 56, 58, 59, 60, 78, 79, 104, 105, 107, 116, 117, 119
adaptive, 1, 5, 9, 12, 21, 27, 30, 31, 39, 48, 51, 52, 55, 65, 72, 85, 94, 96, 98, 99, 100, 110, 115, 118, 127, 128, 129
adaptive control, 98, 118
additive manufacturing, 40, 44
additive manufacturing processes, 40
address, 31, 119
adjustments, 63, 99, 103, 104, 105, 106, 107, 117
adjusts, 30
advanced, 85, 86, 99, 106
advancements, 35, 69, 85, 86, 99, 100, 103, 104, 106, 107, 116, 118, 120
aerospace, 56, 71, 104, 117, 126
agents, 19, 26, 34, 39
AI, 2, 4, 12, 17, 19, 21, 23, 24, 60, 125
AI-powered, 22
Alan Turing, 17, 19
Albert Einstein, 10
Alexa, 21
algorithm, 30, 34, 63

algorithms, 4, 9, 12, 13, 17, 22, 26, 27, 28, 30, 31, 33, 34, 35, 38, 40, 44, 46, 48, 51, 56, 58, 60, 62, 65, 69, 71, 74, 76, 78, 80, 81, 82, 84, 85, 86, 90, 91, 98, 99, 102, 103, 104, 105, 106, 109, 110, 115, 116, 117, 118, 119, 128, 129, 130, 131
align, 51
alignment, 5, 29, 39, 63, 78, 80, 92, 93, 96, 105, 107, 109, 111, 117
Amazon, 21
amplification, 10, 11, 63, 76, 78, 79, 94, 104, 109, 110, 127
amplifier, 5, 56, 62, 65, 76, 78, 93, 96, 105
amplifiers, 5, 62, 75, 76, 79, 111
analysis, 5, 13, 15, 27, 31, 33, 35, 41, 42, 44, 45, 48, 51, 52, 56, 59, 60, 73, 84, 87, 88, 91, 93, 98, 103, 111, 117, 120, 126, 127
analysis methods, 98
analytical, 47, 84
analytics, 85, 86, 116, 119
analyze, 48, 51, 62, 85, 86, 98, 100, 102, 104, 106, 118, 119
analyzing, 30, 63, 85, 88, 105, 106, 116
anomalies, 48, 80, 81, 86, 98, 102, 119, 129
anticipating, 104
applications, 1, 4, 6, 8, 9, 11, 12, 13, 14, 15, 18, 21, 23, 26, 27, 28, 30, 31, 32, 34, 35, 39, 40, 42, 44, 45, 46, 48, 51, 52, 55, 56, 57, 58, 60, 63, 69, 71, 72, 74, 80, 81, 84, 85, 86, 88, 91, 98, 99, 100, 101, 102, 103, 104, 105, 107, 109, 110, 113, 115, 116, 117, 118, 119, 120, 126, 127, 128, 129, 130, 152

approach, 30, 50, 99, 102, 105, 118
approaches, 33, 36, 85, 98, 104, 107, 117
architecture, 100
areas, 117
Arthur Leonard Schawlow, 10
artificial intelligence, 4, 33, 34, 125
artificial intelligence (AI), 4
assemblies, 49
assessing, 49
astrophysical, 57
atmospheric, 100
attosecond, 55, 57
augmented reality, 100, 101
automatically, 52, 53, 63, 98
automating, 61
automation, 12, 15, 26, 34, 49, 51, 52, 53, 98, 99, 100, 102, 110
Automation and Control, 51, 52
automotive, 13, 56, 71, 104, 116, 126
autonomous, 48, 91, 101
autonomous vehicles, 19, 22, 41, 46, 91, 127
average, 35

B

backpropagation, 18
bandwidth, 13, 63, 99
Bayes' Theorem, 29
beam, 1, 4, 9, 11, 28, 29, 31, 34, 38, 39, 41, 44, 51, 55, 56, 57, 58, 60, 62, 63, 76, 77, 78, 79, 80, 81, 82, 84, 92, 94, 99, 104, 105, 109, 110, 111, 117, 118, 119, 124, 125, 126, 130
beam quality, 3, 4, 11, 29, 31, 38, 39, 56, 62, 78, 81, 99, 104, 109, 110, 117, 126, 130
beams, 9, 41, 55, 58, 124, 126
behavior, 51, 63, 85, 99
biochemical, 73
biological processes, 73
biomedical, 40, 44, 46, 48, 118

block diagram, 8, 9, 14, 27, 33, 35, 42, 62, 66, 71, 73, 82, 86, 88, 101
boundaries, 2, 21, 40, 55, 129
Brillouin, 28
British, 17

C

Calculus and Optimization, 27, 29
cancer, 63, 100, 106
capture, 5, 32, 34, 46, 87, 115
case studies, 69, 86, 91
cavity, 8, 9, 29, 38, 94, 96, 151
cellular, 73
center wavelength, 76
century, 4, 10, 18
challenging, 91, 115, 117
chapter, 30, 69
characteristics, 5, 8, 32, 38, 39, 46, 63, 65, 67, 76, 78, 91, 94, 102, 106, 116, 129
characterization, 118, 124
Charles Townes, 10
chemical, 9, 13
Chirped Pulse Amplification, 3, 55, 75, 126, 127
choreography, 14
classification, 31, 35, 48
closed-loop, 39, 65, 67
cluster centers, 32
clusters, 30, 32
coding, 3, 109, 111, 113
coefficients, 31
cognitive, 21, 23
coherence, 9, 38, 110, 111
Coherence, 48
coherent beam, 8
collaboration, 100, 106, 115, 117
combining elements, 33
communication networks, 106
communication technologies, 99
compact, 99, 104
compensation, 62
complex, 1, 4, 17, 22, 24, 29, 30, 31, 32, 38, 39, 45, 46, 48, 57, 66,

71, 73, 76, 78, 85, 102, 104, 105, 109, 110, 117, 119, 129, 130, 152
complex data, 117
complex datasets, 30
complexity, 34, 115
component, 31, 33, 51, 63, 85, 102
components, 5, 8, 9, 27, 28, 32, 39, 40, 46, 49, 55, 60, 62, 75, 77, 78, 80, 82, 92, 93, 96, 105, 110, 126, 130, 131
compositions, 88, 102
comprehensive, 120
compressor, 62, 66, 75, 78, 79, 93, 96, 105
compressor gratings, 62
computational, 4, 38, 71, 99, 130
computational models, 99
computer, 17, 18, 46, 48, 119
computer vision, 46, 48
Computing Machinery and Intelligence, 123
conditions, 51, 52, 61, 65, 67, 85, 91, 98, 100, 103, 105, 115, 118, 119
configurations, 38, 45, 63, 80, 81, 94, 118, 130
consistency, 102, 105, 107
consistent, 51, 105, 119
constant, 63
context, 34, 51, 99
continuous learning, 90, 104
continuously, 53, 65, 86, 100, 104, 106
control, 5, 6, 22, 31, 38, 39, 40, 50, 51, 52, 53, 55, 56, 62, 65, 67, 76, 78, 79, 87, 92, 99, 103, 104, 109, 110, 125, 126, 127, 128, 129
control systems, 51, 52, 53
controlled, 69
conventional, 5, 44, 102, 117
converge, 35
convergence, 4, 12, 23, 68, 115, 116, 119

convolutional, 35, 39
Convolutional Neural Network (CNN), 79
Convolutional Neural Networks (CNNs), 73, 129
cooling systems, 39, 76, 92
core, 8
correction, 13, 41, 100
corrective, 12, 80, 86, 126
costly, 116
costs, 50, 85, 86, 99, 106, 115
covariance matrix, 32
CPA, 3, 11, 55, 56, 58, 61, 62, 63, 65, 66, 75, 76, 78, 93, 96, 104, 105, 107, 116, 117, 126, 127, 128
critical, 5, 22, 26, 28, 29, 34, 38, 40, 46, 56, 71, 75, 78, 80, 91, 92, 93, 102, 109, 119, 130, 131, 152
crystals, 5
cultivate, 116
culture, 52
current, 9, 19, 85, 94, 102
customization, 3, 106, 110, 111, 113
cutting, 2, 4, 8, 11, 13, 14, 15, 22, 28, 29, 31, 34, 51, 56, 60, 69, 70, 85, 86, 92, 98, 100, 103, 104, 105, 106, 110, 118, 119, 123, 126, 128, 129
cybersecurity, 4

D

damage, 13, 40, 57, 77, 105
data, 1, 4, 9, 13, 15, 17, 21, 23, 26, 27, 28, 30, 31, 33, 34, 35, 36, 38, 39, 40, 42, 44, 46, 48, 50, 51, 52, 56, 57, 58, 59, 60, 62, 65, 66, 71, 74, 76, 80, 84, 85, 86, 87, 88, 89, 91, 92, 93, 96, 98, 99, 100, 101, 102, 103, 104, 105, 106, 107, 109, 111, 115, 116, 117, 118, 120, 123, 125, 126, 127, 128, 129, 130, 131

data collection, 26, 27, 46, 59, 84, 87, 91, 92, 96, 116
data flow, 52
data loss, 41
data points, 31, 85, 129
data transmission, 75
database, 67, 88
data-driven, 3, 23, 38, 76, 85, 86, 87, 88, 89, 91, 92, 93, 96, 101, 102, 103, 104, 105, 106, 107, 123, 125, 128
Data-driven
data-driven, 86, 105, 124
dataset, 31, 34
datasets, 4, 18, 23, 26, 28, 30, 31, 32, 38, 44, 51, 56, 57, 73, 81, 84, 85, 90, 91, 100, 102, 115, 118, 119, 129, 130
day-to-day, 85
day-to-day operations, 82, 92, 93
decision trees, 18, 34, 35, 128
decision-making, 48
decisions, 30, 34, 117
decomposition, 28
deep learning, 5, 18, 19, 31, 71, 73, 79, 110, 129
Deep Learning, 122
deeper insights, 73, 111
defects, 102, 104
defense, 55, 58, 60
deformable mirrors, 39
degradation, 5, 39, 63, 92
dermatology, 119
descent, 29, 35, 36
design, iv, 3, 6, 27, 30, 38, 39, 85, 96, 99, 104, 109, 118, 123, 124, 130, 151, 152
detect, 52, 86, 100, 117
detection, 4, 5, 21, 26, 28, 39, 46, 48, 57, 92, 100, 101, 127, 128, 129, 130
detectors, 39
developed, 61, 87

development, 4, 9, 10, 11, 12, 13, 19, 21, 30, 38, 39, 41, 46, 60, 73, 75, 84, 85, 88, 90, 91, 99, 102, 103, 106, 107, 110, 111, 115, 117, 118
developments, 2, 11, 19, 101, 104, 106, 116
deviations, 100
diagnostic, 3, 12, 13, 41, 48, 56, 63, 65, 71, 106
diagnostics, 34, 48, 63, 65, 86, 98, 99, 100, 115, 117, 118, 119
diagram, 1, 46, 48, 50, 52, 58, 60, 65, 70, 71, 75, 78, 87, 91, 92, 103, 111, 113
differential calculus, 29
differentiation, 46
dimensionality, 28, 30, 32, 33, 65, 130
dimensionality reduction technique, 28, 130
directed-energy weapons, 58
directional, 8, 9
diseases, 13, 48
disk, 104, 105, 107, 116, 117
Disk lasers, 107
dispersion, 62, 94, 105, 117
dispersive element, 62
displays, 14
disrupt, 104
disruptions, 117
dissipation, 105
distance, 90
distances, 32
distortion, 105, 117
distortions, 29, 39, 76, 79, 109
divergence, 9
diverse, 90, 115, 118, 119
domains, 115, 117
doped, 104
downstream, 47, 94
downtime, 5, 13, 39, 50, 51, 53, 63, 81, 84, 100, 105, 106, 115, 116, 128, 131

drifts, 62
drive, 12, 27, 46, 117, 119
drones, 58
drug, 12, 23, 73
dye, 8, 11
dynamic, 14, 40, 61, 71, 73, 78, 82, 101, 103, 104, 118
dynamically, 52, 62, 65, 100, 105, 106, 117, 119
dynamics, 4

E

early-stage, 48
educational, 116, 117, 119
effective, 35, 51, 115, 118
Effective feature extraction, 34
effectiveness, 86, 103, 106, 119
efficacy, 41, 58
efficiency, 3, 8, 9, 11, 13, 15, 22, 26, 29, 31, 33, 38, 39, 40, 44, 45, 50, 51, 53, 56, 57, 58, 60, 66, 69, 70, 71, 78, 81, 82, 84, 85, 86, 88, 92, 94, 96, 98, 99, 100, 102, 104, 105, 107, 115, 116, 117, 119, 120, 126, 130
efficient, 51, 52, 63, 85, 99, 100, 101, 103, 107, 116, 117
efficiently, 85
eigenvalue, 28
eigenvalues, 28, 32
eigenvectors, 28, 32
electrical, 8
electron, 8
electronics, 8, 11, 19
electrons, 8
elements, 8, 53, 63, 104, 126
ELI-NP, iii, 125
emission, 10, 12
empirical, 4, 38
enables, 52, 67, 102, 104, 105, 106, 117, 118, 119
energy, 31, 62, 63, 99, 132
energy source, 8, 9
engineering, 4, 10, 11, 34, 44, 149
engineers, 31, 116, 117, 119, 123

engraving, 13, 22, 29, 56, 92, 100, 106, 119
ensemble methods, 18, 130
entertainment, 14, 15, 100
environment, 22, 26, 30, 33, 39, 48, 53, 79, 127, 129
environmental, 48, 52, 62, 65, 67, 85, 91, 98, 105, 106, 116, 118, 119
environments, 5, 19, 41, 46, 48, 52, 66, 84, 90, 98, 103
equipment, 50, 51, 52, 53, 115, 117
era, 101
error, 5, 13, 19, 29, 35, 38, 41, 44, 76, 106, 117
error correction, 106, 117
errors, 100
essential, 33, 34, 52, 62, 65, 99, 100, 104, 105, 106, 107, 115, 116
ever-changing, 53
evolution, 85, 106
evolving, 35, 118
example, 31, 33, 52, 65, 106, 118
excel, 31
excitation, 99
excite, 8
exotic semiconductors, 99
experience, 24, 109, 120
experimental, 5, 38, 85, 89, 102, 109, 111, 117, 151
experimental setups, 85, 117
experiments, 4, 5, 22, 30, 44, 57, 60, 84, 85, 89, 117, 120
expert, 5
exploration, 17, 38, 41, 44, 46, 53, 60, 71, 111, 151
exponential growth, 18
eye, 12, 22, 48, 55, 57, 63, 126

F

factors, 62
failures or anomalies, 50
fault, 4, 5, 100, 101
fault detection, 100

feedback, 26, 27, 30, 33, 34, 39, 48, 49, 50, 52, 59, 63, 65, 67, 76, 78, 79, 81, 82, 84, 87, 89, 91, 92, 94, 96, 98, 99, 102, 103, 110, 127, 128, 129, 132, 151
feedback loop, 51
femtosecond, 62
femtosecond pulses, 4
fiber lasers, 5, 8, 55, 56, 58, 104, 126
Fiber-optic communication, 13
fields, 1, 4, 5, 9, 12, 14, 15, 17, 21, 23, 26, 30, 40, 44, 46, 55, 57, 60, 63, 73, 74, 84, 91, 99, 100, 115, 116, 117, 118, 120, 126
finance, 4, 19, 21
fine-tune, 63, 86
fine-tuning, 62
flexibility, 105
flow, 48, 52
flowchart, 58, 60, 65, 92, 96
fluctuations, 41, 76, 79, 81, 85, 92, 93, 106
fluorescence, 15, 73
focal spot size, 63
focus, 51, 118
forecasting, 39
formula, 35
fostering, 52, 106, 116
foundation, 52
framework, 28, 33, 44, 45, 46, 96
free-space optical communication, 39, 74
frequency drift, 94
FSO communication systems, 41
fundamental, 2, 12, 27, 34, 41, 45, 57, 76, 81, 122, 151
fundamentals, 9
future, 3, 9, 10, 12, 17, 22, 45, 53, 60, 85, 87, 98, 99, 100, 102, 103, 106, 107, 111, 116, 117, 118, 119, 152

G

gain media, 88, 102, 104

gain medium, 8, 9, 38, 62, 94, 105, 126, 127
gain saturation, 62
gaining traction, 33
gas, 8, 11, 151
generalization, 34
generation, 8, 56, 57, 78, 107, 116, 151
Generative Adversarial Networks, 118
global, 104, 106
Google, 18, 109, 111
gradient, 18, 29, 35
graph, 12, 19, 23, 107
grating, 62, 66, 78, 79, 94, 96
grating separation, 78, 79
grating stretcher, 62

H

health, 40, 111
healthcare, 13
heavy-duty, 105
helium-neon, 11
HHG, 85
hierarchical, 35, 129
hierarchically, 31
high power, 62, 104, 125
higher energy, 8
higher energy state, 8
higher harmonics, 57
highlights, 1, 9, 12, 15, 19, 28, 53, 60, 61, 70, 71, 87, 107, 112
high-power, 105, 107
High-Power Laser, 122
high-powered lasers, 11
historical, 30, 50, 51, 67, 85, 86, 98
human, 48, 51
human-like, 21, 23
hybrid, 3, 111, 112, 118
Hybrid, 33
hypothesis, 28

I

ideal, 62
identification, 102
illustrate, 69, 86, 91

illustrated, 33
illustrates, 35, 52, 107
image, 35, 49, 50
imaging, 15, 22, 31, 32, 34, 39, 41, 44, 46, 48, 56, 73, 98, 115, 129
imaging modalities, 48
impact, 12, 34, 38, 40, 53, 56, 58, 60, 63, 69, 79, 84, 86, 92, 94, 100, 119
implement, 48, 53, 80, 86, 109
implications, 91
improve, 34, 35, 51, 85, 98, 100, 118, 125
industrial, 49, 51, 69, 85, 86, 103, 105, 107, 115, 118, 119
Industrial, 98
industries, 1, 4, 12, 13, 18, 19, 21, 24, 50, 53, 55, 70, 71, 98, 100, 103, 104, 105, 106, 115, 116, 117, 126
industry, 117, 119
inefficiencies, 102, 106
influence, 8, 9
information, 65, 106
innovation, 4, 9, 11, 12, 35, 41, 45, 46, 53, 84, 85, 111, 117, 120
innovations, 2, 4, 11, 30, 55, 74, 86, 91, 98, 99, 101, 102, 104, 117
innovative, 115, 116
insights, 88, 100, 103, 104, 105, 106, 117, 119, 132
instability, 80, 92, 93
Institutions, 116
instrumental, 11, 31, 38, 105
Integrating, 6, 39, 40, 46, 52, 60, 61, 63, 69, 78, 86, 92, 119
integration, 1, 9, 11, 14, 19, 27, 42, 44, 46, 52, 53, 55, 56, 58, 60, 61, 63, 65, 86, 98, 99, 100, 101, 103, 105, 107, 110, 112, 116, 117, 118, 119, 120, 127
intelligent, 51, 87, 98, 103

intensity, 4, 22, 28, 30, 31, 34, 40, 44, 51, 55, 56, 57, 58, 60, 62, 71, 77, 80, 84, 86, 106, 109, 116, 130
interacting, 30, 33
interactions, 5, 11, 30, 31, 32, 44, 45, 46, 73, 78, 111, 118
interdisciplinary, 106, 115, 116, 117
interference, 99
Internet, 13
interplay, 53, 105
interpret, 13, 22, 46, 48, 57, 117
interpretation, 47, 48, 117
interrelationship, 4
intersection, 33, 115, 118
intervention, 51, 63
interventions, 88, 119
Introduction to Machine Learning, 26, 122
invaluable, 30, 104, 132
invention, 10
inventions, 4
inversion, 28, 151
iterative, 29, 90

J

journey, 132

K

Kerr lensing, 63
key role, 100, 101
K-means, 31
k-means clustering, 31, 33
K-means clustering, 31

L

labeled datasets, 26
laboratories, 5, 149
landscape, 35, 53, 115, 116, 118
large dataset, 71, 79
laser, 1, 4, 5, 8, 9, 10, 11, 12, 13, 14, 15, 19, 21, 26, 27, 28, 30, 31, 33, 34, 35, 38, 39, 40, 42, 44, 45, 46, 48, 49, 51, 52, 53, 55, 56, 57, 58, 60, 61, 62, 63, 65, 69, 70, 71,

73, 74, 76, 78, 79, 80, 81, 82, 84, 85, 86, 87, 88, 90, 91, 92, 93, 96, 98, 99,100, 101, 102, 103, 104, 105, 106, 107, 109, 110, 111, 113, 115, 116, 117, 118, 119, 120, 123, 124, 125, 126, 127, 128, 129, 130, 131, 149, 151
laser cutting, 70
laser designs, 85
laser frequencies, 57
Laser Fundamentals, 122
laser light, 106, 117
laser machining, 30, 123
laser operation, 104
laser operations, 51, 53, 86, 98, 99
Laser Physics, 122
laser systems, 53, 86, 98, 101, 104, 116, 118
laser technology, 1, 4, 9, 10, 11, 12, 13, 14, 15, 19, 21, 26, 27, 29, 30, 35, 38, 40, 45, 46, 51, 56, 57, 58, 60, 71, 81, 84, 86, 98, 101, 102, 103, 106, 107, 110, 111, 115, 116, 118, 126, 127, 128, 129
laser therapy, 98
laser-assisted, 87
laser-based, 34, 48, 90, 100, 105, 106, 115, 119
laser-driven, 1, 15
laser-induced fluorescence, 13, 73
lasers, 48, 52, 63, 69, 85, 88, 98, 99, 100, 102, 103, 104, 105, 106, 107, 116, 117, 119, 122, 124, 125
Lasers, 12, 46, 60, 99, 100, 102, 122, 123, 124
LASIK, 13, 15
learning, 30, 31, 33, 34, 35, 48, 51, 52, 63, 65, 67, 69, 85, 86, 87, 90, 98, 99, 100, 101, 103, 104, 105, 106, 107, 115, 116, 117, 118, 119, 124
leverage, 14, 27, 33, 39, 98, 106, 111, 112, 115, 116, 119

LiDAR (Light Detection And Ranging), 46
lifespan, 51, 100, 103, 105, 117
light, 8, 9, 10, 11, 14, 39, 58, 60, 127, 151, 152
Light, iii, 8, 149
light amplification, 8
likelihood, 28, 46, 130
Linear Algebra, 27, 28
linear model, 31
Linear regression, 31
liquid, 8
long distances, 106
lower energy, 8
lower energy state, 8

M

machine, 30, 33, 34, 35, 43, 48, 50, 51, 52, 53, 61, 62, 63, 65, 66, 69, 85, 86, 87, 88, 98, 99, 100, 102, 103, 104, 105, 106, 107, 115, 116, 117, 118, 119, 120, 123, 124, 125
machine learning, 1, 4, 5, 9, 12, 13, 14, 17, 19, 21, 23, 24, 26, 27, 28, 30, 33, 34, 35, 38, 39, 40, 42, 44, 45, 46, 48, 50, 51, 52, 53, 55, 56, 57, 58, 59, 60, 61, 62, 63, 65, 66, 69, 70, 71, 73, 74, 76, 78, 80, 81, 82, 84, 85, 86, 87, 88, 90, 91, 92, 93, 96, 98, 99, 100, 101, 102, 103, 104, 105, 106, 107, 109, 110, 111, 113, 115, 116, 117, 118, 119, 120, 123, 124, 125, 126, 127, 128, 129, 130
Machine learning, 34, 48, 51, 62, 65, 67, 100, 102, 103, 104, 105, 106, 116, 119, 123
machine learning (ML), 4, 17
machine learning models, 65, 118
machine perception, 48
machines, 13, 17, 21, 31, 46
maintain, 59, 62, 67, 76, 79, 81, 94, 96, 99, 104, 106, 119
maintaining, 63

maintenance, 1, 5, 13, 38, 42, 50, 51, 53, 60, 63, 67, 84, 92, 96, 100, 101, 104, 106, 107, 109, 115, 116, 125, 128, 130, 131
malfunctions, 99, 117
manage, 51, 105
management, 107
manual, 5, 40, 62, 80, 92, 96, 105
manufacturing, 1, 4, 6, 9, 11, 13, 15, 22, 31, 38, 40, 44, 45, 46, 50, 51, 55, 56, 58, 60, 71, 84, 86, 98, 100, 102, 103, 104, 105, 106, 107, 110, 115, 116, 118, 119, 126, 129, 130
map, 30
mapping, 14, 46, 48, 128
marks, 6, 12, 119
maser, 10
material, 1, 5, 11, 13, 15, 22, 28, 30, 31, 33, 34, 35, 40, 44, 45, 46, 55, 56, 60, 69, 70, 71, 85, 86, 88, 90, 98, 99, 102, 103, 104, 105, 116, 117, 118, 119, 129
material processing, 1, 6, 11, 31, 44, 55, 60, 104, 118
material properties, 31, 33, 34, 85, 119
material science, 14, 45
material strength, 44
materials, 8, 12, 13, 22, 27, 32, 41, 44, 45, 46, 55, 58, 60, 71, 84, 88, 99, 100, 102, 103, 104, 105, 118, 126, 129
mathematical, 27, 30, 32, 34, 109
mathematical expression, 32
mathematician, 17
MATLAB, 109
matrices, 28
matrix, 28
maximum variance, 32
Mean Squared Error (MSE), 29
measurable property, 34
measurements, 44, 46, 57, 79, 85, 90, 91, 100

medical, 2, 4, 5, 9, 11, 12, 15, 19, 21, 27, 32, 39, 40, 48, 51, 55, 58, 60, 63, 65, 70, 71, 84, 85, 86, 87, 88, 98, 99, 100, 102, 103, 104, 106, 117, 118, 119, 126, 128
Medical, 123
medical diagnostics, 21, 39, 99, 102
medicine, 48, 73, 98, 100, 106
methodologies, 30, 34, 86, 115
methods, 33, 40, 62, 85, 102, 116, 124
microelectronics, 56
micromachining, 63, 65, 105
micro-machining, 55
microwaves, 10
milestone, 119
misalignment, 5, 29, 80, 92, 131
misalignments, 78, 80, 94
missiles, 58
mitigate, 105
model, 30, 34, 35, 36, 62, 67, 86, 125
model evaluation, 26, 27
model training, 26, 27, 47, 59, 109, 127
modeling, 4, 23, 28, 29, 44, 45, 56, 85, 102, 104, 111, 115, 128
mode-locking, 93
models, 4, 5, 17, 22, 26, 27, 28, 31, 33, 34, 35, 38, 39, 40, 44, 45, 46, 59, 62, 65, 67, 69, 71, 73, 76, 81, 84, 85, 88, 89, 90, 91, 92, 94, 96, 106, 109, 110, 115, 118, 127, 128, 129, 130
modern, 8, 104, 106
Modern, 104, 116
modulators, 39
monitor, 22, 51, 57, 63, 76, 81, 86, 94, 104
monitoring, 3, 5, 22, 39, 53, 57, 58, 60, 63, 65, 76, 82, 85, 86, 87, 91, 93, 96, 98, 99, 100, 101, 103, 104, 111, 118, 127, 130

monochromatic, 8, 9
morphology, 87
MRI, 48
multi-pass, 5, 62, 94, 110, 111
multiple, 35, 51, 62, 120
multiplication, 28
nanotechnology, 14
navigate, 35, 46, 118
navigation, 48
Netflix, 18, 21
neural network, 66, 124
neural networks, 5, 18, 31, 33, 34, 35, 38, 39, 41, 45, 65, 76, 98, 118, 129

N

Neural networks, 31, 63, 129
neurons, 31
no-code, 3, 109, 110, 111, 112, 113
noise, 62, 91
nonlinear, 11, 63, 65, 79, 110
non-linear, 4
nonlinear effects, 63, 65
normalized, 67
novel, 88, 99
nuclear fusion, 57

O

object, 48, 128, 129
object detection, 48
object recognition, 48
obstacle detection, 48
obstacles, 116
OCT, 48
oncology, 48, 57
operation, 3, 5, 6, 9, 10, 27, 28, 39, 60, 82, 85, 92, 96, 105
operational, 50, 53, 66, 85, 86, 99, 100, 104, 105, 106, 107, 116
ophthalmology, 12, 48, 119, 126
optical, 5, 8, 9, 11, 28, 29, 39, 41, 44, 55, 62, 63, 74, 77, 80, 99, 104, 110, 126, 152
optical components, 62
optical wireless communication, 99

optics, 2, 4, 11, 30, 31, 39, 77, 85, 94, 109, 111
optimal, 19, 26, 29, 31, 38, 40, 44, 52, 62, 63, 65, 66, 67, 69, 70, 76, 79, 81, 87, 94, 96, 99, 102, 103, 104, 105, 106, 118, 128, 129, 152
optimal boundaries, 31
optimal performance, 62
optimization, 3, 4, 5, 6, 29, 31, 33, 38, 40, 42, 46, 51, 52, 56, 59, 61, 65, 70, 76, 78, 80, 82, 91, 93, 96, 101, 102, 103, 107, 117, 127, 128, 130
optimize, 9, 12, 22, 26, 29, 33, 35, 40, 44, 51, 56, 57, 62, 63, 65, 67, 69, 70, 74, 76, 82, 85, 86, 93, 96, 98, 100, 104, 110, 115, 116, 117, 118, 119, 129, 130
optimizing, 30, 34, 35, 65, 100, 104, 105, 106
Optimizing, 4, 75, 80
organization, 33
oscillator, 62, 75, 76, 79, 93, 96, 105
outcomes, 30, 31, 34, 35, 51, 85, 98, 100, 104, 106, 115, 117, 119
output, 5, 8, 11, 29, 30, 34, 38, 39, 55, 56, 60, 62, 76, 78, 79, 81, 82, 92, 94, 100, 102, 105, 110, 116, 119, 126, 130
overview, 9, 127
Overview, vii, 7
parallel, 104
parameter, 63
parameters, 4, 5, 22, 26, 28, 30, 31, 33, 34, 35, 38, 39, 40, 42, 44, 45, 51, 52, 55, 56, 57, 59, 60, 62, 63, 65, 67, 70, 71, 76, 78, 79, 80, 81, 82, 84, 85, 86, 87, 90, 92, 94, 96, 98, 100, 103, 104, 105, 106, 115, 116, 117, 118, 119, 127, 128, 129, 130

P

Partial derivatives, 29
particle acceleration, 57
patient, 40, 49, 56, 58, 71, 87, 98, 106, 117, 119, 128
patient-specific, 87
pattern recognition, 17, 23, 35, 126
patterns, 4, 5, 14, 17, 21, 26, 27, 30, 31, 33, 34, 39, 40, 44, 48, 51, 73, 80, 81, 84, 85, 87, 93, 98, 102, 109, 117, 119, 127, 129, 131
peak, 62, 86, 105
Perceptron, 17
performance, 2, 5, 9, 11, 12, 18, 22, 26, 27, 28, 30, 33, 34, 35, 38, 39, 40, 41, 44, 51, 53, 59, 60, 61, 63, 65, 67, 74, 76, 78, 80, 81, 82, 84, 86, 88, 91, 92, 93, 96, 98, 99, 100, 101, 102, 103, 104, 105, 106, 115, 126, 129, 130, 152
performance metrics, 31, 51, 65, 88, 102
phase, 9, 34, 38, 45, 57, 62, 65, 67, 71, 79, 88, 94, 96
phenomenon, 8, 34
photonics, 4, 6, 8, 32, 123
phototherapy, 12
photothermal therapies, 40
physical, 4, 123
physicists, 119
physics, 4, 10, 45, 55, 57, 109, 122, 124
pie chart, 15
plasma, 85, 124
plasma generation, 56
plays, 103, 116
pollution, 100
potential, 5, 12, 14, 17, 21, 26, 39, 45, 51, 81, 87, 93, 103, 105, 106, 116, 117, 128, 131
power, 2, 4, 8, 18, 23, 29, 31, 33, 34, 38, 40, 44, 51, 55, 56, 57, 58, 60, 62, 69, 70, 76, 78, 80, 81, 82, 86, 91, 92, 93, 96, 102, 103, 104, 105, 107, 109, 110, 116, 117, 119, 126, 127, 129, 130
powerful, 3, 13, 18, 31, 99, 106, 111, 129, 152
practical, 69
precise, 31, 35, 62, 91, 100, 104, 105, 116, 118
precision, 2, 4, 5, 11, 13, 15, 22, 29, 31, 33, 34, 35, 38, 39, 40, 44, 48, 49, 51, 52, 55, 56, 57, 58, 60, 63, 70, 80, 81, 86, 92, 94, 98, 99, 100, 102, 103, 104, 106, 107, 115, 116, 117, 118, 119, 126, 128
pre-configured, 110
predict, 31, 33, 34, 35, 50, 51, 63, 65, 66, 85, 88, 98, 99, 100, 106, 117
predicting, 30, 31, 34, 62
predictions, 4, 21, 26, 27, 28, 34, 35, 45, 67, 91, 102, 130
predictive, 1, 4, 5, 13, 26, 38, 40, 42, 44, 45, 50, 51, 53, 56, 60, 63, 65, 69, 70, 84, 85, 86, 91, 96, 98, 100, 101, 102, 104, 107, 116, 118, 125, 127, 128, 130, 131
predictive modeling, 44, 46
preprocessed, 51, 65, 67
preventative, 5
primary, 51, 85
Principal Component Analysis, 32
Principal Component Analysis (PCA), 28, 130
principle, 8, 10
proactive, 53, 116
Probability and Statistics, 27, 28
problems, 35, 123
process, 8, 32, 34, 50, 51, 62, 65, 85, 102, 123
processing images, 50
production, 39, 50, 51, 56, 57, 69, 70, 102, 104, 106
productivity, 86, 99, 105, 106, 116, 117

professionals, 1, 13, 71, 106, 109, 110, 111, 118, 152
programming, 4, 21, 109, 110, 126
progression, 23, 73, 101, 107
projection, 14
projects, 120
proliferation, 18
prominent, 11, 85, 86
properties, 5, 8, 9, 11, 13, 14, 15, 28, 44, 56, 88, 99, 102, 104
protocols, 41, 52, 53
prototype, 11
pulse, 4, 11, 31, 34, 39, 40, 44, 55, 56, 60, 61, 62, 63, 65, 67, 75, 76, 78, 80, 82, 84, 93, 96, 104, 105, 107, 109, 110, 117, 119, 123, 127, 130, 149, 151
pulse compression, 62, 65, 67
pulse compressor, 62
pulse duration, 31, 34, 36, 62, 63, 66, 118, 119
pulse shape, 63, 76, 80
pulse shaping, 5, 61
pulse width, 4, 95, 105, 117
pulses, 62
pump, 62
pump source, 9
pumping configuration, 38
pumping mechanism, 8
Python, 109, 111
PyTorch, 109

Q

quality, 31, 35, 50, 51, 52, 98, 104, 115, 116, 119
quantum, 10, 11, 22, 99, 100, 102
quantum computing, 99, 100
quantum dot lasers, 99
Quantum machine learning, 23
quantum mechanics, 10
quantum particles, 22

R

Radiation, 8
Raman, 13, 15, 28, 152
Raman spectroscopy, 13, 15

random error, 31
random forests, 35, 36
ranging, 100
re-acquisition, 51
reactions, 9
real-time, 2, 5, 9, 21, 26, 27, 38, 39, 40, 44, 46, 48, 50, 51, 52, 56, 57, 58, 59, 62, 65, 66, 71, 76, 78, 81, 82, 85, 86, 87, 91, 92, 93, 96, 98, 99, 100, 101, 102, 103, 104, 105, 106, 107, 110, 111, 115, 117, 118, 119, 120, 127, 128, 129
real-world, 34, 118
recompressed, 62
Recurrent Neural Network (RNN), 79
refine, 98
regenerative, 5, 62
regulate, 51
reinforcement, 30, 33, 34, 36, 63, 65, 124
Reinforcement
 reinforcement, 26, 30, 34, 39, 41, 78, 127, 129
reinforcement learning, 2, 19, 22, 26, 33, 34, 36, 39, 40, 63, 65, 71, 76, 79, 81, 110
Reinforcement learning, 30, 34
relationships, 35, 102
reliability, 5, 11, 26, 34, 41, 53, 58, 74, 82, 84, 86, 91, 102, 131
remarkable, 4, 19, 71, 74
research, 2, 4, 5, 9, 11, 12, 13, 15, 17, 21, 23, 24, 41, 55, 57, 60, 69, 73, 74, 75, 81, 84, 85, 86, 92, 105, 110, 115, 118, 119, 149
researchers, 31, 98, 116, 117, 118, 119
reshaping, 107, 118
resonator, 8
revolutionize, 88, 98
reward signal, 31
robotics, 41, 91, 100, 101

robust, 33, 44, 45, 79, 84, 96, 98, 118, 130
ruby laser, 4, 11

S

safety, 51, 53, 98, 104, 106
scattering, 74
scientific, 2, 4, 5, 11, 23, 24, 41, 55, 57, 60, 105, 117, 118
scientists, 98, 115, 117, 119
seed pulse, 62
self-correct, 100
self-focusing, 105, 117
self-phase modulation, 63
semiconductor, 8, 11, 99, 101
semiconductor lasers, 8, 11
sensor, 21, 39, 91, 129, 130
sensors, 18, 65, 91, 99, 101, 103, 115, 130
series, 65
settings, 33, 51, 52, 62, 65, 86, 100
setups, 5, 29, 57, 110
shaping, 106
shut down, 52, 53
signal, 13, 74, 94, 106, 117
signal processing, 13, 74, 106
significant, 5, 6, 10, 11, 18, 19, 23, 26, 27, 28, 41, 44, 49, 55, 69, 85, 86, 98, 99, 102, 115, 129
Silfvast, 122
simulate, 118
simulations, 4, 71, 85, 152
Siri, 21
skin, 123
skin treatments, 57
smarter, 1, 6, 9, 27, 63, 71, 81, 100, 101, 102, 107
smart-systems, 98
smoothly, 93
social media, 18
solid-state, 8, 11, 99, 101, 104, 105, 107, 110, 116, 117
Solid-state lasers, 105, 107
source, 62, 124, 132
spatial, 28, 39, 41, 46, 48

spatial coordinates, 28
spatial resolution, 48
spectral, 32, 62, 63, 67, 76, 78
spectral broadening, 32
spectral width, 62, 76, 79
spectroscopy, 28, 32, 41, 63, 100
speed, 1, 13, 15, 22, 39, 40, 86, 104, 106, 129
SPM, 63
stability, 66, 102, 104, 107
statistical, 85
steepest descent, 35
stimulated emission, 8
Stimulated Emission, 8
strategy, 30
streamline, 87, 111
strength, 132
stretch, 4, 95
stretched pulse, 62
stretcher, 62, 75, 78, 79, 93, 96, 105
structural, 102
structural integrity, 13, 40
subset, 34
suitable, 30, 65, 105
sum, 32
supervised, 30, 31, 33, 34, 36, 65
supervised learning, 22, 26, 34, 38, 41, 69, 81
Supervised learning
supervised learning, 26
support vector machines, 18, 31, 33, 34, 39
surgery, 4, 12, 22, 31, 55, 56, 63, 87, 100, 102, 103, 126
surgical procedures, 87, 106
surgical settings, 40
surveillance, 58
Svelto, 122
synergy, 9, 14, 41, 46, 48, 56, 85, 86, 98, 99, 103, 120
system, iv, 1, 4, 5, 6, 22, 23, 26, 27, 29, 31, 34, 38, 39, 46, 48, 52, 61, 62, 63, 65, 66, 71, 73, 74, 76, 78,

81, 87, 91, 92, 93, 96, 100, 103, 104, 105, 109, 110, 111, 125, 127, 128, 129, 130, 131, 151, 152
systems, 1, 4, 5, 6, 9, 11, 12, 13, 21, 24, 26, 27, 28, 30, 32, 38, 39, 40, 42, 46, 50, 51, 52, 53, 55, 56, 57, 58, 60, 62, 63, 69, 73, 74, 75, 76, 78, 81, 84, 86, 91, 92, 93, 98, 99, 100, 101, 102, 103, 104, 105, 106, 107, 109, 110, 115, 116, 117, 118, 119, 120, 123, 126, 127, 129, 130, 152

T

target, 13, 91, 95
techniques, 3, 5, 11, 13, 15, 21, 26, 29, 33, 38, 40, 44, 46, 48, 49, 51, 57, 63, 65, 69, 73, 81, 84, 86, 89, 98, 99, 115, 116, 119, 123, 128
technological, 4, 18, 24, 53, 55, 91
technology, 8, 30, 31, 33, 34, 35, 48, 49, 51, 52, 53, 85, 86, 91, 98, 99, 100, 103, 104, 106, 107, 115, 116, 117, 118, 119, 120, 122, 123, 149
telecommunications, 4, 9, 11, 12, 13, 15, 22, 74, 84, 99, 103, 104, 106, 117
temperature, 41, 80, 85, 92, 93, 109
temporal, 62, 67, 79
temporal coherence, 62
TensorFlow, 109, 122
terahertz, 11
terminology, 33, 35
testing, 34, 36, 67, 152
Theodore Maiman, 11
theoretical, 10, 152
theories, 10
therapeutic, 40, 56
thermal effects, 31, 38, 62, 94, 104, 109
thermal expansion, 78
thermal lensing, 94, 105, 109, 110

thin, 105
threat neutralization, 58
throughput, 40, 74, 106, 119
time, 30, 33, 48, 53, 63, 65, 67, 86, 103, 105, 107, 115, 116, 117
time mapping, 48
timeline, 12, 19, 107
Ti-sapphire, 3, 5, 61, 62, 63, 65, 66, 93, 96
tissue, 31, 40, 48, 57, 63, 87
tissues, 48
tolerance, 39
tools, 65, 100, 106
traditional, 3, 4, 14, 40, 44, 58, 73, 84, 85, 98, 102, 111, 118
trained, 5, 18, 33, 34, 48, 62, 65, 67, 71, 79, 81, 88, 91, 94, 96, 127
training, 18, 26, 28, 30, 34, 36, 38, 73, 76, 85, 115
training data, 34
trajectory, 12, 23
transfer, 13
transform, 100
transformation matrix, 28
transformative, 4, 12, 21, 30, 45, 46, 56, 69, 74, 81, 91, 103, 104, 106, 115, 118, 120
transmission, 13, 15, 22, 41, 75, 99, 106
treatment, 63, 106, 119, 123
treatments, 2, 5, 11, 15, 40, 55, 56, 58, 60, 70, 71, 85, 98, 106, 117, 118, 126, 128
trial and error, 34
trial-and-error, 85
tuning, 5, 40, 55, 57, 62, 65, 67, 76, 82, 86, 124, 152
turbulence, 39, 41, 74
tweak, 94

U

ultrafast, 2, 4, 11, 31, 32, 39, 55, 57, 111, 126, 127
ultrafast lasers, 4

146

ultra-intense, 5, 6
ultrashort, 61, 62, 149
Ultrashort Pulse Laser
  Technology, 122
uncorrelated, 32
unexpected, 50, 117
unlabeled data, 30, 34
unprecedented, 6, 18, 23, 55, 78, 85, 86, 104, 107, 116
unsafe, 52, 53
unsupervised, 30, 31, 33, 34, 36
unsupervised learning, 30, 34
  unsupervised learning, 22, 26, 28, 129
use, 35, 52, 53, 88, 107, 116

V

valuable, 105
variability, 115
variables, 31, 52, 105, 118
vectors, 28
versatile, 103
vibration, 94

virtual assistants, 21
visualization, 32, 73, 117

W

wakefield, 85
waste, 51, 52, 86, 98, 100, 104, 119
wavefront, 39
wavelength, 4, 9, 31, 34, 36, 40, 55, 60, 62, 76, 79, 84, 102, 118
wavelengths
  wavelength, 8, 32
waves, 9
welding, 4, 8, 11, 13, 15, 22, 28, 30, 51, 56, 60, 71, 85, 98, 100, 104, 105, 106, 118, 126, 128, 129
winter, 17
workflows, 2, 58, 102, 110, 111, 112
workforce, 106
world, 107

Y

years, 12, 19, 107

ABOUT THE AUTHOR

Laser physicist/engineer educator specializing in ultrashort pulse laser technology, working at Extreme Light Infrastructure-Nuclear Physics facility, Romania. Ph.D. from the Korea Advanced Institute of Science and Technology, South Korea. Postdoctoral research at Instituto Superior Técnico, Portugal. He was an assistant professor at King Saud University, Saudi Arabia. As an associate professor, he has also served at COMSATS  University Islamabad and Lahore University of Management Sciences, Pakistan. He taught, researched, and established laser laboratories at various universities. His research interests are related to laser engineering and technology. He is a recipient of several research fellowships.
tayyabmrn@gmail.com
http://tayyabimran.weebly.com

BOOKS BY THIS AUTHOR

Laser Technology Made Easy: Building a Sealed Tube Carbon Dioxide (CO2) Laser System from Scratch for Students and Hobbyists (amazon.com)
https://a.co/d/9ohchJB

This short book focuses on developing a sealed tube carbon dioxide (CO2) laser from scratch with a reasonable depth without attempting to be overly broad, a mix of theory and experimental aspects. We start by looking at the fundamental concepts of laser, like population inversion, pumping scheme, feedback system, and active medium, and further move to the working of the CO2 laser, how the CO2 laser is electrically pumped and stabilized using the homemade cooling system.

It is helpful for readers who want to work on developing lasers with limited resources and is useful for graduate and undergraduate students taking a course on lasers and light sources. Hobbyists who want to work on developing lasers and have a basic understanding of the laser may get the most out of this book.

Guide to Building a LASER: Air-based Nitrogen Laser Construction (amazon.com)
https://a.co/d/hMehOCr

Guide to Building a LASER is a meticulously crafted resource that offers a clear and concise exploration of TEA nitrogen lasers. This book combines theory and practicality to provide a well-rounded understanding of the subject. Join us on this illuminating journey, where each chapter is thoughtfully designed to contribute to your deeper comprehension of the marvels of laser science and technology. Unveiling the marvels of air as a lasing medium and employing the Blumlein cavity configuration, this guide caters to seasoned scientists, aspiring students, and curious minds alike. Beginning with the fundamental principles of gas and TEA nitrogen lasers, the book intricately delves into the Blumlein cavity configuration, its design, and operational nuances. It demystifies high-voltage pulse ge neration and provides a detailed roadmap for inexpensively constructing a TEA nitrogen laser system. Offering insights into testing, tuning, and practical applications, the guide ensures optimal performance, accompanied by thorough safety guidelines. Aspiring researchers, engineers, and enthusiasts will find this book a guiding light into the fascinating world of TEA nitrogen lasers, merging theory, practical insights, and real-world applications. Embark on a journey into the future of laser science and technology with this compelling and professional resource.

A Practical Handbook: Spectrometer and Interferometer Design with ZEMAX OpticStudio (amazon.com)
https://a.co/d/1W2JXXU

The book is a practical guide aimed at students, researchers, and professionals interested in optical system design. It bridges the gap between theoretical concepts and real-world applications, using ZEMAX OpticStudio as a powerful tool for designing, simulating, and optimizing optical systems. The book focuses on the design and simulation of spectrometers and interferometers, two critical optical instruments.

The book is divided into several chapters. It begins with the foundational knowledge of spectrometry and interferometry, followed by an introduction to ZEMAX software and its key features. The book's core delves into step-by-step simulations of different types of spectrometers, including optical and Raman spectrometers and interferometers like the Mach-Zehnder and Michelson interferometers. The simulations guide the reader through the design process, from conceptual design to simulation results, providing hands-on tutorials and real-world examples. The authors also highlight their previous publications on ZEMAX-based design simulations of spectrometers and interferometers, which have contributed to shaping the content of this book. This comprehensive resource aims to make the complex task of optical system design more accessible through practical applications and tutorials.

www.ingramcontent.com/pod-product-compliance
Lightning Source LLC
Chambersburg PA
CBHW071406210526
45465CB00001B/274